Fossil News

The Journal of Avocational Paleontology

Vol. 23.2 & 23.3 — Summer/Fall 2020

FRONT COVER: *Titanoboa* surfacing from the muddy waters of a swamp to grab a crocodilian snack. *Titanoboa* was a very large snake that lived during the Middle-Late Paleocene in what is now northeastern Colombia. The only known species is *Titanoboa cerrejonensis*, the largest snake ever discovered. Based on fossilized vertebrae, researchers estimated that the largest individuals reached a total length of around forty-two feet and weighted more than a ton. Some researchers believe that *Titanoboa* was predominantly a fish-eater. If true, the trait is unique among boas. (Copic Marker and colored pencil on Schoellershammer paper. 30 cm x 21 cm.) **BACK COVER:** *Cryopterygius*, an ichthyosaur known from the Late Jurassic of Central Spitsbergen, Norway. *Cryopterygius* is believed to have reached a length of around sixteen feet. The Slottsmøya Member of the Agardhfjellet Formation is a Lagerstätte known for producing ichthyosaur fossils. Images © Esther van Hulsen; used by permission.

CONTRIBUTORS TO THIS ISSUE

ALBERT J. COPLEY (writer) is Assistant Professor Emeritus at Truman State University in Kirksville, Missouri, and has taught widely across the West and Midwest. His Cinema Expeditions, Inc. produces and sells videos and 35-mm slide sets for geological education, and he has written some dozen ebooks dedicated to such topics as the geology of caverns, the Sonoran Desert National Monument, and the petroglyphs of New Mexico, Utah, and Arizona.

ESTHER VAN HULSEN (featured artist) received her degree in Communication Design and Scientific Illustration from the Academy of Arts Minerva in Groningen, Holland, in 2004, and has lived in Norway since 2006. Shortly after art school, she began working as a full-time wildlife artist and, since 2010, has been a frequent collaborator with paleontologist Jørn Hurum of the Natural History Museum at the University of Oslo, publishing books on marine reptiles, eurypterids, and the early-primate fossil, "Ida." She works as a professional wildlife illustrator and paleoartist.

MASSIMO PIETROBON (cartographer & writer) was born in Italy but today lives in Barcelona, Spain. He has also lived in Brazil, Morocco, Sierra Leone, and Angola. Trained as an architect, he has expanded his commitment to the arts over the years, becoming an illustrator, graphic designer, model maker, cartographer, and an official interpreter in six different languages. Through art and his creative impulse, he has always sought ways to express the world that lives inside him.

MATS E. ERIKSSON (writer) is a professor of paleontology at Lund University, Sweden, and a metal aficionado. He studies creatures from the Paleozoic Era and tries to reconstruct and understand their interrelationship with the surrounding physical environment. He's the author of *Another Primordial Day: The Paleo Metal Diaries* as well as of numerous scientific and popular-science articles. He's profoundly committed to scientific outreach and, among writing and other activities, is the co-creator of the traveling educational exhibit, *Rock Fossil on Tour*. He lives in Malmö, Sweden.

PAUL D. TAYLOR (writer) is an invertebrate paleontologist specializing in bryozoans, and his research includes their taxonomy, biomineralogy, and paleoecology. He spent more than thirty years at the Natural History Museum of London and headed the Invertebrate and Plants division between 1990 and 2003. He is the author of *A History of Life in 100 Fossils* and *Fossil Invertebrates*, among many other publications. He retired from the museum in 2018 but continues there in a part-time postdoc research project.

SPENCER G. LUCAS (book reviewer) is a stratigrapher and paleontologist who has been Chief Curator of Geology and Paleontology at the New Mexico Museum of Natural History and Science (Albuquerque, NM) since 2009. His research focuses on biostratigraphic problems of the late Paleozoic, Mesozoic, and early Cenozoic, and he has undertaken extensive field research in the American West, Kazakhstan, China, Mexico, Nicaragua, and Costa Rica. He is the author of seven books and more than a thousand scientific articles.

Fossil News: The Journal of Avocational Paleontology is published quarterly. Unless otherwise indicated, all content is ©*Fossil News: The Journal of Avocational Paleontology* for twelve months following the cover date. Within that period, none of the articles, images, or any other materials contained herein may be reused without the express written permission of *Fossil News*. This embargo includes web and social-media publication of all kinds, though exceptions will gladly be considered. At the expiration of the twelve-month term, all rights reverts to the original creators, and reprint requests should be addressed to them.

HOW TO SUBSCRIBE: Domestic US rates: $50.00 USD for four quarterly issues, payable via PayPal or Zelle (both via FossilNews@FourCatsPress.com) or Venmo (@FourCatsPress). Outside the U.S., please contact us for a quote. Single issues: $15.00 USD single numbers; $20.00 USD double numbers (plus postage if outside the US). Complete info at: tinyurl.com/fnsubscribe.

DISPLAY ADVERTISING: Subscribers receive 10% off display advertising. Our current "Rate Sheet & Specifications" are online at: tinyurl.com/fnadvertise.

EDITOR/PUBLISHER: Wendell Ricketts
CONTRIBUTING EDITORS: Diana Fattori & Nando Musmarra EDITOR EMERITA: Lynne Clos
FOUNDING EDITOR: Joe Small

SUBMISSION GUIDELINES: *Fossil News* welcomes submissions from all those with an interest in paleontology and fossil collecting. Submissions and queries to: FossilNews@FourCatsPress.com.

HOW TO SUBMIT YOUR WORK: Please read our guidelines (tinyurl.com/fnguidelines) or contact us for details.

BOOK REVIEWS: *Fossil News* does not accept unsolicited or previously published reviews. It's always best to query first. If possible, send clips of your published work.

EDITING AND REVISION: *Fossil News* reserves the right to edit submissions for length, grammar, punctuation, structure, and style.

PAYMENT: We offer contributors a complimentary copy of the issue in which their work appears and a choice of (a) two insertions of a display ad (up to one-half page) or (b) 50% off an annual or gift subscription.

Fossil News
The Journal of Avocational Paleontology
FossilNews@FourCatsPress.com
www.FossilNews.org

Summer/Fall 2020: 978-1-7348050-8-6

ESTHER VAN HULSEN

A PORTFOLIO OF RECENT WORK

Above: Styracosaurus, named for a "spike at the end of a spear" plus "sauros" (lizard), was a herbivorous Cretaceous ceratopsian dinosaur. Its nose horn may have grown as long as two feet. The function of the frill has been the subject of many theories over the years. The first partial *Styracosaurus* was found in Canada in 1913 by C. M. Sternberg, and the famous American paleontologist, Barnum Brown (named after the circus entrepreneur, P. T. Barnum, but no relation), found another nearly complete skeleton in 1915 on one of his many expeditions for the American Museum of Natural History. Twenty-two years after Sternberg's find, a crew from the Royal Ontario Museum returned to the same site and managed to recover most of the rest of the skeleton. (*Copic marker and colored pencil on Schoellershammer paper. 21 cm x 30 cm.*)

Ophthalmothule cryostea was a plesiosaur that lived around the Jurassic-Cretaceous boundary and is known from the Slottsmøya Member Lagerstätte of the Agardhfjellet Formation in Spitsbergen, the largest and only permanently populated island of the Svalbard archipelago in northern Norway. The type specimen was discovered by Dr. Aubrey Roberts. *Ophthalmothule* belonged to a family of medium-sized plesiosaurs and was notable for its enormous eyes. *(Copic marker and colored pencil on Fabriano paper. 42 cm x 30 cm.)*

Above: An undescribed ichthyosaur from the Jurassic Posidonia Shale of Holzmaden, Germany. The drawing is based on a fossil that probably represented a juvenile because the head was so large compared to its body. *(Acrylics on art board, 30 cm x 40 cm.)*
Right: A Triassic scene from Spitsbergen Island in the Svalbard Archipelago, Norway. Recently, the mountains of Spitsbergen have yielded the fossils of many marine animals, several of which are reconstructed in this scene. *(Copic marker and colored pencil on Fabriano paper. 30cm x 40 cm.)*

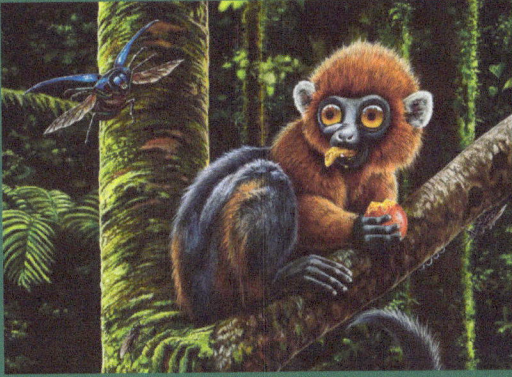

*F*ossil News introduced readers to the work of artist Esther van Hulsen in Summer 2016 when we featured van Hulsen's remarkable reconstruction of *Keuppia*, a ninety-five-million-year-old octopod. Van Hulsen drew with ink reconstituted from the fossilized ink sac of a specimen from Lebanon. Since then, her wide-ranging work has continued, including wildlife images, scientific illustrations, and nature-inspired paintings, notecards, and other art. She has illustrated numerous books dedicated to re-imagining the lives and environments of ancient vertebrates and invertebrates, including, with Jørn Hurum, the story of the discovery and study of "Ida," the only known specimen of the Eocene basal primate, *Darwinius masillae* (above, left), found in the Messel pit not far from Frankfurt, Germany. More of her work appears on her Facebook page and on her website, esthervanhulsen.com.

Above left: One of van Hulsen's illustrations from *Ida: The Extraordinary History of a 47-Million-Year-Old Primate*, originally published in Norwegian and since translated into French, Chinese, Russian, Dutch, and Danish. (One day, we hope, it will also appear in English.) *Ida* was chosen as the Best Scientific Children's Book in France in 2014. *Above right:* van Hulsen at work on her image of *Keuppia*. *Below: Caudipteryx* feeding its young. *Caudipteryx* was a peacock-sized early Cretaceous theropod dinosaur whose fossils were first discovered in China in 1997. *(Copic marker and colored pencil on Fabriano paper. 21 cm x 30 cm.)*

Anhanguera, a
pterodactyloid
pterosaur from the
Early Cretaceous of
Brazil. Its name comes
from the word *añanga* in
the language of the Tupi, one of the largest groups of
indigenous people in Brazil prior to colonization, and
refers to "the spirit protector of the animals." Early Jesuit
missionaries, however, employed *añanga* as their word
for the Devil. *Anhanguera* was a piscivore (fish-eater) with
a wingspan of up to fifteen feet. *(Copic marker and colored
pencil on Schoellershammer paper. 30 cm x 21 cm).*

A fuzzy *Deinonychus*, a dromaeosaurid dinosaur (or "raptor," the group most closely related to modern birds) that lived during the early Cretaceous in what is now the American West. Though clear fossil evidence of bird-like feathers has not yet been found for *Deinonychus*, paleontologists consider it highly likely that the "terrible claw" had feathers as did its dromaeosaur relatives. *(Copic marker and colored pencil on Schoellershammer paper. 30 cm x 21 cm.)*

Above: An undescribed species of pterosaur, flying reptiles that lived from the late Triassic to the end of the Cretaceous. Though many pterosaurs were apparently fish eaters, some hunted land animals, ate insects or fruit, or cannibalized other pterosaurs. *(Watercolor and colored pencil on art board. 30 cm x 42 cm.)* *Below*: An early Cretaceous raptor, *Utahraptor*, taking a nap. *(Both: Copic marker and colored pencil on Fabriano paper. 21 cm x 30 cm.)* *Facing page: Smilodon* is ready for a closeup. *(Copic marker and colored pencil on Schoellershammer paper. 30 cm x 21 cm.)*

An Esther van Hulsen menagerie: *Bottom spread:* A pair of *Smilodon* stalk a mastodon mired in an asphalt lake at what is today the La Brea Tar Pits in Los Angeles, California. The Tar Pits Museum reports having recovered at least fifteen mastodons and more than 2,000 sabre-toothed cats in excavations over the years, making them the second most common animal found there. *Left, top: Stegosaurus*, an herbivorous armor-plated dinosaur from the Late Jurassic. *Allosaurus* and *Ceratosaurus* may have been its two biggest predators. *Right, top: Carnotaurus*, a Late Cretaceous dinosaur known from South America. The thick horns above its eyes are unusual among carnivorous dinosaurs. *Middle: Psittacosaurus*, a Late Cretaceous dinosaur known

mainly from Asia, Mongolia, and Siberia. *Psittacosaurus* is thought to be a basal ceratopsian and is found so commonly that it has become one of the best known dinosaur genera. *Bottom: Mei long*, an Early Cretaceous duck-sized theropod dinosaur known from China. *Mei long* means "dragon sleeping soundly," and here is an individual doing just that.

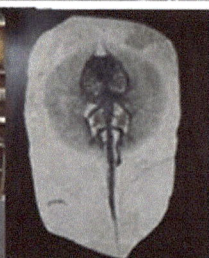

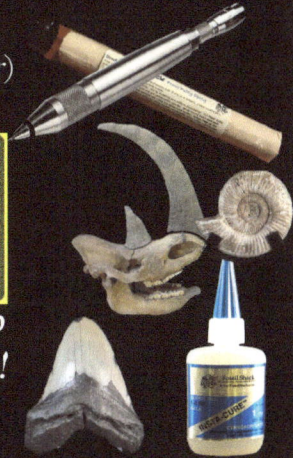

In 2016, four colleagues in science—friends from graduate school at the University of Colorado-Boulder—created 500WomenScientists, a grassroots organization built on a commitment to speak up for science and for marginalized communities in science. Four years later, more than 20,000 women in STEMM (science, technology, engineering, mathematics, and medicine) have signed the 500WomenScientists pledge (see sidebar), committing themselves to building an inclusive scientific community capable of training a more diverse group of future leaders in science. As of the organization's 2019 annual report, 388 "pods" had been created around the world (local chapters of women scientists who meet regularly to develop a support network, make strategic plans, and take action).

With partners that include the Alan Alda Center for Communicating Science, the National Council for Science and the Environment, 500QueerScientists, the Earth Science Women's Network, and the Union of Concerned Scientists, 500WomenScientists has launched numerous projects to carry their vision forward around the U.S. and the world. A complete list is on their site, but here are a few highlights:

- **Inclusive Scientific Meetings:** Scientific meetings can be invigorating, promote the exchange of ideas, foster new collaborations, and provide opportunities to reconnect with existing colleagues. But not all scientists have positive experiences at scientific meetings: some feel isolated or left out (intentionally or otherwise); some encounter barriers such as lack of childcare or safe bathroom spaces; and some are targets of harassment and assault. 500WomenScientists published a guide to planning, conducting, and assessing workshops, symposia, seminar series, and conferences that increase diversity, equity, and inclusion.

- **Guides to Writing Op-Eds**: Editorials and articles that focus on women in science, science funding, and other science issues relevant to local communities are an important part of organizing and public education. So far, members have published on such topics as why women leave science careers, tax plans that endanger future scientists, and legislation aimed at dismantling science-based policy-making, and their work has appeared in *Scientific American, The Atlanta Journal-Constitution, The Durango Herald* (Colorado), *The Knoxville News Sentinel* (Tennessee), and *The Hill* (Washington, D.C.), among other publications.

- **Wikipedia Edit-A-Thons:** Less than a fifth of English-language biographies on Wikipedia tell the stories of women. In 2019, pods edited nearly 200 Wikipedia entries to increase representation of women in STEMM online. To celebrate International Women's Day last March, 500WomenScientists launched a global campaign of Edit-A-Thons.

- **Sci-Mom Journeys**: 500WomenScientists proposed and promotes a series of hard-hitting policy positions on key areas that affect mothers and aspiring mothers in STEMM.

- **Request a Woman in STEMM.** Launched in January 2018, the project provides opportunities for members of the media, scientific colleagues, conference organizers, educators, and others to find and include more women and underrepresented identities in their work. Thirteen thousand individuals from 140 countries are now included in the network (see map). Through the searchable directory on the 500WomenScientists website (https://500womenscientists.org/request-a-scientist), anyone who needs scientific expertise or consultation can search by name, discipline, or geographical area, and find direct contact information (there are currently 348 women in geology in the directory and sixty-seven with a specialization in paleontology, paleoecology, paleobotany, or paleoclimatology).

500WomenScientists says the group's values are rooted in: "Recognizing that science touches the lives of every person on this planet; advocating for a strong role of science in society; identifying and acknowledging structural inequities and biases in science; pushing for equality and standing up to inequality, discrimination, and aggression; pushing to develop and strengthen access for traditionally underrepresented groups to fully participate in and become leaders in science; supporting the education and careers of all scientists; enhancing scientific mentorship and encouraging an atmosphere of collaboration; stepping outside of our research disciplines to communicate our science and engage with the public; and using the language and wonder of science to bridge the divides that separate societies and to enhance global diplomacy.

The Pledge
(published in *Scientific American* on November 17, 2016)

In this new era of anti-science and misinformation, we as women scientists re-affirm our commitment to build a more inclusive society and scientific enterprise. We reject the hateful rhetoric that [has] targeted minority groups, women, LGBTQIA, immigrants, and people with disabilities, and attempted to discredit the role of science in our society. Many of us feel personally threatened by this divisive and destructive rhetoric and have turned to each other for understanding, strength, and a path forward. We are members of racial, ethnic, and religious minority groups. We are immigrants. We are people with disabilities. We are LGBTQIA. We are scientists. We are women.

- As women scientists, we are in the position to take action to increase diversity in science and other disciplines. We resolve to continue our pursuits with renewed passion and to find innovative solutions to the problems we face in the U.S. and abroad. Together, we pledge to:
- Identify and acknowledge structural inequalities and biases that affect the potential of all individuals to fulfill their goals;
- Push for equality and stand up to inequality, discrimination, and aggression;
- Push to strengthen the support for traditionally under-represented groups to fully participate in and become leaders in science;
- Support the education and careers of all scientists;
- Step outside of our research disciplines to communicate our science and engage with the public;
- Use every day as an opportunity to demonstrate to young girls and women that they are welcome and needed in science;
- Set examples through mentorship and through fostering an atmosphere of encouragement and collaboration, not one of divisiveness;
- Use the language of science to bridge the divides that separate societies and to enhance global diplomacy.

Today, we invite the women in science and our colleagues to declare our support to each other and to all minorities, immigrants, people with disabilities, and LGBTQIA. Our scientific work may be global, yet we will take action in our own communities and we will work towards an inclusive society, where science and knowledge can be embraced and everyone has the opportunity to reach their potential.

As women in science, technology, engineering and math, as role models to young girls and women, as leaders in our communities, we accept this challenge.

With new members arriving daily, the "Request a Woman in STEMM" project is growing all the time. This map, which shows the number of contacts in each part of the world, is updated regularly on the 500WomenScientists site.

The Folklore of Fossil Molluscs

Devil's Toenails, Bulls' Hearts, Horses' Heads, and Thunderbolts

Paul D. Taylor

Molluscs are the second-largest phylum of invertebrate animals after the Arthropoda—both in terms of the number of species living today and in their extremely rich fossil record. The majority of mollusc species possess a resistant calcareous shell (usually but not always external) that accounts for often excellent preservation. Most fossil molluscs belong to one of three major groups: bivalves (oysters, clams, and so on), gastropods (snails and slugs), and cephalopods (ammonites, belemnites, etc.). The Mollusca, however, also includes lesser-known groups such as the monopla-

cophorans (thought to be extinct until 1952, when living relatives were discovered), the polyplacophorans (chitins), the aplacophorans (small, deep-water marine animals that are today usually shell-less, though fossil forms possessed valves), rostroconchs (which date to the Early Cambrian but disappeared at the end of the Permian), scaphopods (tusk shells), and the Helcionelloida (an extinct and ancient group whose affinities remained the subject of debate until they were assigned to a class of molluscs in 1991).

Fossil molluscs are usually recognizable because of their close similarities to the shells of familiar species of modern molluscs. Some, however, are not quite so straightforward, and these are more likely to have given rise to fanciful stories about their origins and significance. In the context of the early history of paleontology and of doubts about the origin of fossils, some more obscure ancient molluscs were dubbed "difficult fossils" by British geologist, historian, and academic, Martin Rudwick. They include the solid internal casts (steinkerns) formed by lithification of sediment enclosed within a shell after the defining shell itself is subsequently lost. In addition, there are some mollusc fossils—notably belemnite guards—that bear little resemblance to any living species, adding to their enigmatic nature.

Belemnites: Thunderbolts and Devil's Fingers

The first fossils I ever came across were belemnites from the Jurassic Kellaways Rock, collected while on a boyhood adventure following the course of the disused Hull and Barnsley Railway that ran at one point alongside South Cave Station Quarry. I took them home not knowing what they were but my father, who, like many of his generation had not been educated about fossils, told me they were "thunderbolts" hurled to the ground during thunderstorms. The streamlined, bullet- or missile-like shape of belemnite guards seemed consistent with this idea (indeed, the name belemnite is derived from the Greek belemnon, meaning dart or javelin) and,

Five belemnite guards oriented as though they were projectiles flung from the sky like "thunderbolts." *Belemnitella minor* from the Cretaceous Paramoudra Chalk of Norfolk, England.

to add credence to the notion of "thunderbolts," the heavy rains associated with thunderstorms do occasionally wash away the topsoil and bring fossils such as belemnites to the surface, almost as though they had come from the sky.

According to an East Anglian horseman, "As the sun draws up water, so the clouds draw up substance from the earth—sulfur and so on. When there's a clap o'thunder, down all this comes as thunderbolts" (quoted in Evans, 1966: 131).

The striking shapes of belemnite guards, coupled with a robustness that imparted a high potential for fossilization,

Longitudinal section of the Early Jurassic belemnite *Acrocoelites* showing the chambered phragmocone (slightly crushed and displaced to the right) above the guard.

have led to a plethora of folkloric names in addition to thunderbolts. In some regions of England, they are known as Fairies' Fingers, Devil's Fingers, or Saint Peter's Fingers (Bassett, 1982; Duffin & Davidson, 2011). The earliest mention of belemnites in Scotland dates to 1703, when Scottish writer Martin Martin (Scottish Gaelic: Màrtainn MacGilleMhàrtainn) referred to them as "botstones."

In German folklore, belemnites are known by numerous different names, including Alpschoß (nightmare shot), Fingerstein (finger stone), Gespensterkerze (ghostly candle), and Katzenkegel (cat's skittle—the name of the pins in the bowling game of ninepins) (Hegele, 1997). Scandinavian folklore envisages belemnites as candles belonging to elves, gnomes, or pixies, hence the Swedish name Vetteljus (Gnomes' Lights). They were believed to protect unchristened children from being transformed into changelings by trolls (Duffin, 2008). In Chinese folklore, belemnites are known as sword stones (劍石 or jiàn shí).

There are several records of belemnites recovered from archaeological sites. Oakley (1974) described a Bronze Age burial site in Yorkshire where a belemnite was found with a female skeleton, seemingly a testament to the cult status of these fossils. Fragments of altered, amber-colored belemnites with fine perforations, which may have been used as charms, were found at a 20,000-year-old archaeological site known as Kootenai 17 on the Don River in Russia (Boriskovskii, 1956). In ancient Egyptian hieroglyphic inscriptions, objects resembling belemnites appear in references to the Middle Kingdom god Min, a deity of virility. According to Newberry (1910), fossil belemnites and certain arrowheads were cult-objects in the Egyptian Middle Kingdom, representing thunderbolts or, possibly, phallic symbols and, by association, Min.

Belemnites were once believed to have medicinal qualities, and were used as cures for both rheumatism and sore eyes in humans and horses. The treatment for horses involved crushing the fossils into a dust that would then be blown into the eyes of the animal. In Scotland, they were steeped in water and employed to cure horses of the worms that caused distemper (Oakley, 1974). They were also used to keep a person from being struck by lightning or bewitched by demons from the sky (Kennedy, 1976).

But belemnites, of course, are in fact the internal shells of an extinct group of squid-like cephalopods with hooks rather than suckers on their arms. Their true affinity is unclear from the bullet-shaped guard alone—the coarse, radiating calcite crystals forming the belemnite guard could be taken to suggest an inorganic origin like the crystals in a geode. More complete specimens preserve the phragmocone. This chambered shell, which is constructed of the readily dissolved

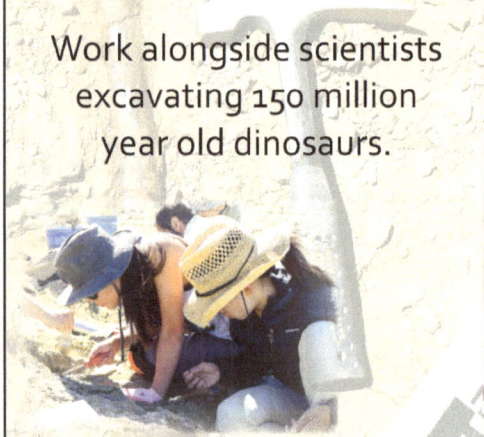

mineral aragonite, fits into a conical depression in the blunt end of the guard and closely resembles the chambered shells of such cephalopods as the living nautilus, providing crucial evidence for the true affinities of these Jurassic and Cretaceous animals.

Pagoda Stones

Longitudinal sections through Paleozoic orthoconic nautiloids are known in China as Pagoda Stones (宝塔石 or bǎotǎ shí) because of their vague resemblance to the tiered towers found in many temples. The abundance of these fossils explains the origin of the name Pagoda Formation for an Ordovician deposit in South China.

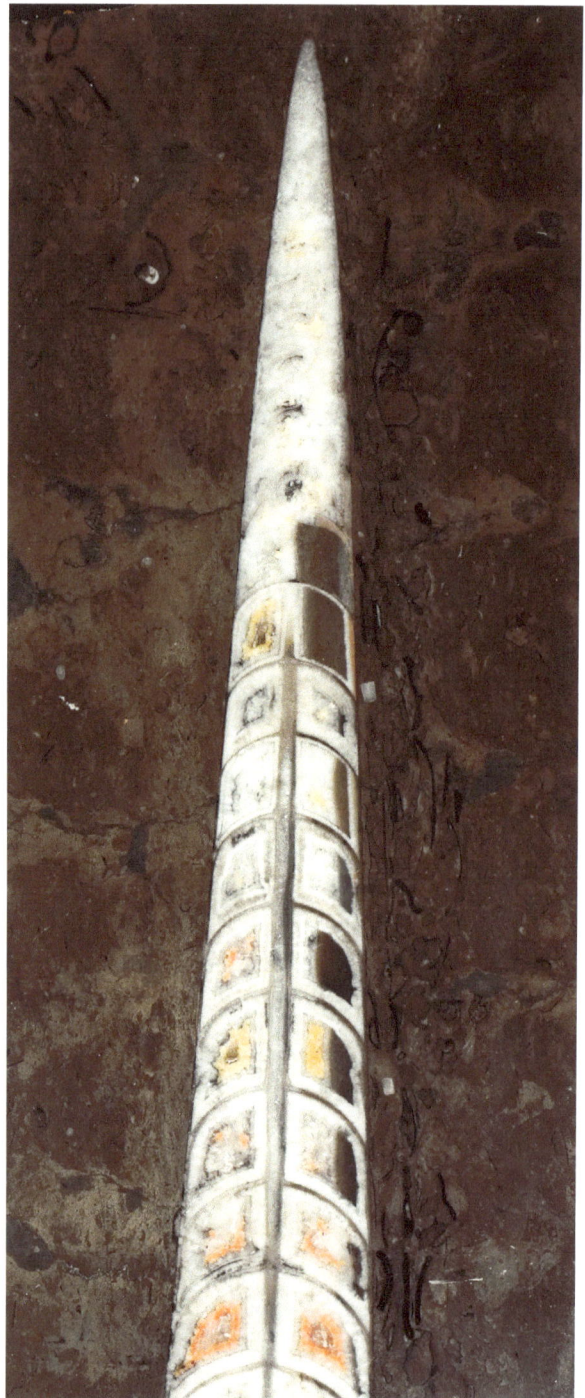

Devil's Toenails

"Devil's toenails," shells of the oyster *Gryphaea*, are among the most abundant fossils found in the British Jurassic. The calcite shell of *Gryphaea* is thick and survives weathering and erosion of the sediments in which it is fossilized. It is also sufficiently resilient to have endured transportation by rivers and Pleistocene glaciers—eroded specimens of *Gryphaea* are often found in river gravels and glacially deposited boulder clays in such regions of England as Suffolk and Gloucestershire.

The robust, curved left valve of *Gryphaea*, marked with prominent growth bands, superficially resembles a thick toenail. It is unclear whether *Gryphaea* shells were once believed to be the actual toenails of devils or simply corresponded to the popular conception of what a devil's toenail ought to look like.

Devil's toenails are particularly common in the Lower Jurassic rocks around Scunthorpe, formerly quarried intensively for the iron ore that was economically important for this Lincolnshire town. They feature in the town's coat of arms, adopted in 1936. Knell (1988) quoted a passage from a diary written on 10 April 1696 by a local man, Abraham de la Pryne, who said that powdered *Gryphaea* was used to cure "a horse's sore back."

In Scotland, fossil *Gryphaea* shells are known in Old Gallic as *clach crubain*, translated as "crouching shell" (Oakley, 1974). They were apparently used in the seventeenth and eighteenth centuries to cure pain in the joints. Oakley made the interesting

Facing page: A "Pagoda Stone," a longitudinally sectioned orthoconic nautiloid from the Ordovician of China. *This page, top*: a "Devil's toenail," the oyster *Gryphaea*. This example comes from the Early Jurassic of Gloucestershire, England. *Right*: The old coat-of-arms of Scunthorpe, Lincolnshire, England, depicting two valves of *Gryphaea* over the chain in the shield. The banner translates the original Latin motto: *Refulget Labores Nostros Coelum*.

THE HEAVENS REFLECT OUR LABOURS

point that their contorted appearance is suggestive of painful joints, an example of sympathetic medicine ("like cures like").

Stone Cores

The fossilized internal molds of bivalves were puzzling objects to early naturalists and common people alike (these molds are often known by the German name "steinkern"). This style of preservation is particularly characteristic of species of bivalves that had aragonite shells. Unlike the more stable calcite shells found, for example, in *Gryphaea*, aragonite is routinely dissolved by water filtering through the rock. In cases in which a shell has been filled by sediment reasonably shortly after death, the shell is not fossilized but the hardened sediment core is. Steinkerns are solid and may fall cleanly out of the rock when it is broken open. The often

bizarre and vexing appearance of bivalve steinkerns led to misconceptions about their origins and spawned the names "'Osses' 'Eds" (that is, "Horses' Heads) and "Bulls' Hearts" by which they are known in folklore. Glørstad, Nakrem, and Tørhaug (2004) described a remarkable steinkern of the Ordovician bivalve *Cyrtodonta* from a Mesolithic site in southeastern Norway that had been collected and subsequently sculpted to emphasize "attributes" that were apparently believed to resemble a human woman.

'Osses' 'Eds

Steinkerns of a group of bivalves common in the Jurassic of southern England, when viewed in a particular orientation, vaguely resemble the heads of miniature horses. The "eyes" of the horse are actually molds of scars left by the muscles that originally closed the valves together, and the "ears" are the pointed beaks of the two valves. The Oxford naturalist Robert Plot (1677) referred to examples of these fossils from Headington near Oxford as "Hippocephaloides," alluding to their horse head-like shape. They are now known by the scientific name *Myophorella hudlestoni*.

Steinkerns of a related trigoniid bivalve (*Myophorella incurva*) are conspicuous in the Portland Stone of Dorset,

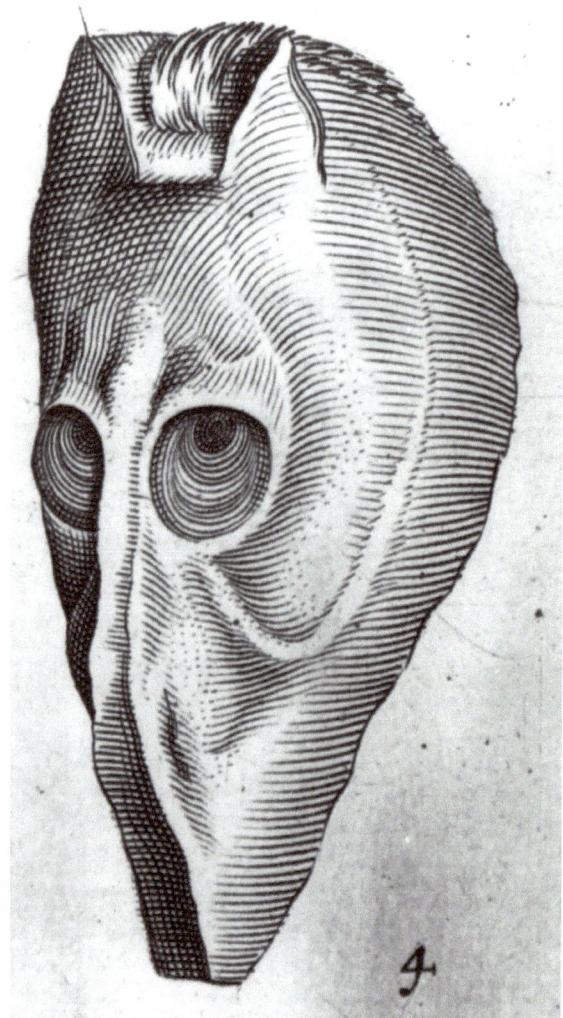

particularly in a level called the "Roach," and are frequently used decoratively as a facing stone on buildings (such as, for example, the Economist Buildings in the City of Westminster). Quarrymen on the Isle of Portland, speaking in local dialect, knew them as 'Osses' 'Eds .

Bulls' Hearts

Another type of bivalve steinkern from the British Jurassic has been called a Bull's Heart (Fig. 8). Like 'Osses' 'Eds, these were found by Robert Plot at Headington, who referred to them as "*Bucardites*." They are now known by the scientific name *Protocardia*. Their resemblance to a heart becomes apparent when the fossils are viewed from the side, with the mold of the left valve on one side and that of the right valve on the other.

Evidence that fossils of the same type have long been known to humankind comes from the discovery of a *Protocardia* steinkern in a Bronze Age barrow (burial mound) at Aldbourne in Wiltshire (Oakley, 1974). This barrow was built on chalk, though the fossil comes from the underlying Upper Greensand and must have been collected and taken to the site of the barrow.

Facing page: The 'Osses' 'Eds of folklore. *Left:* The internal mold (steinkern) of the bivalve *Myophorella incurva* from the Late Jurassic of Portland, Dorset, England. *Right:* Robert Plot's figure of an 'Osses' 'Ed emphasizing the resemblance to a miniature horse's head with eyes, mane, and ears clearly drawn. *This page, top left:* Steinkern of the bivalve *Glossus* from the Pliocene of Italy and, below it, a "Bull's heart," the internal mold of the Jurassic bivalve *Protocardia*, as illustrated by Robert Plot.

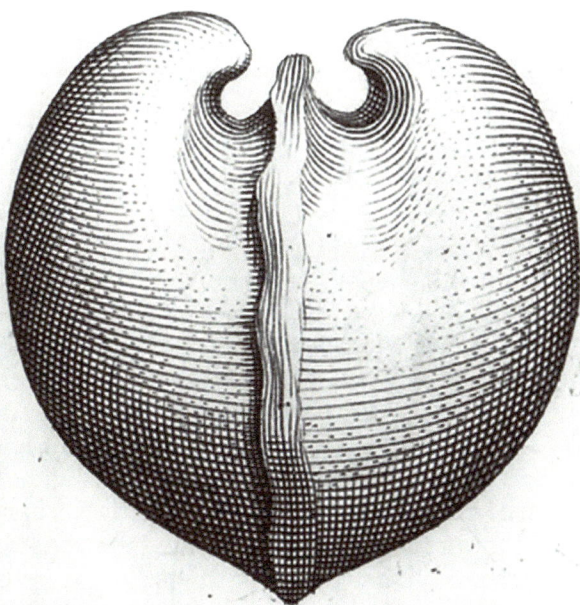

References

Bassett, M.G. (1982). *Formed Stones, Folklore and Fossils.* Cardiff, Wales: National Museum of Wales.

Boriskovskii, P I. (1963). Essays on the Paleolithic of the Don Basin (in Russian). *Materialy i Issledovaniya po Archeologii SSSR*, 121, 80-124.

Duffin, C. (2008). Fossils and Folklore. *Ethical Record*, 113, 17-21.

Duffin, C.J. and Davidson, J.P. (2011). Geology and the Dark Side. *Proceedings of the Geologists' Association, 122*, 7-15.

Evans, G. E. (1966). *Patterns under the Plough: Aspects of the Folk-Life of East Anglia.* London, England: Faber and Faber.

Glørstad, H., Nakrem, H. A., and Tørhaug, V. (2004). Nature in Society: Reflections over a Mesolithic Sculpture of a Fossilised Shell. *Norwegian Archaeological Review*, 37, 95-110.

Hegele, A. (1997). Donnerkeil und Teufeflsfinger: Belemniten in Voldsglauben und Volksmedizin. *Fossilien*, 1/97, 21-26.

Kennedy, C. B. 1976. A Fossil for What Ails You: The Remarkable History of Fossil Medicine. *Fossil Magazine*, 1, 42-57.

Knell, S. J. (1988). *The Natural History of the Frodingham Ironstone.* Scunthorpe, England: Scunthorpe Borough Museum and Art Gallery.

Martin, M. (1703). *A Description of the Western Isles of Scotland.* London, England: Andrew Bell.

Newberry, P. E. (1910). The Egyptian Cult-Object and the "Thunderbolt." *Annals of Archaeology and Anthropology*, 3, 50-52.

Oakley, K. P. (1974). Folklore of Fossils. Part I. *New York Paleontological Society Notes*, 5(1-2), 9-17.

Plot, R. (1677). *The Natural History of Oxfordshire, Being an Essay towards the Natural History of England.* Oxford, England.

DON'T FED ME IN

WHAT THE AAPS KEEPS GETTING WRONG ... AND WHY THE CONVERSATION ABOUT FOSSIL COMMERCE NEEDS TO START OVER AGAIN FROM SCRATCH

WENDELL RICKETTS

As one digs deeper into the national character of the Americans, one sees that they have sought the value of everything in this world only in the answer to this single question: how much money will it bring in? — Alexis de Tocqueville

It is the fundamental contradictatoriness of the United States of America—the illogical but optimistic notion that you can create a union of individuals in which every man is king. — Susan Orlean, The Orchid Thief: A True Story of Beauty and Obsession

The ground is being emptied so that ... things can be possessed. This isn't just about stewardship. It is an obsession that runs the gamut of our desires. — Craig Childs, Finders Keepers: A Tale of Archaeological Plunder and Obsession

The Association of Applied Paleontological Sciences, founded in 1978 as the American Association of Paleontological Suppliers, is a trade organization of less than 200 commercial dealers in fossils and minerals from many parts of the world. They are, according to their home page, "a professional association of commercial fossil dealers, collectors, enthusiasts, and academic paleontologists for the purpose of promoting ethical collecting practices and cooperative liaisons with researchers, instructors, curators and exhibit managers in the paleontological academic and museum community." Literally tens of thousands of other fossil merchants exist online, in physical stores, at pay-to-dig sites, and in other contexts.

According to a July 2020 update of its membership list, nearly 80% of AAPS members are located in the United States. The "dinosaur" states (Arizona, Colorado, South Dakota, Wyoming, Montana, and Utah) are heavily represented among members (46%), and among those who make up the AAPS executive board, administration, and committees (58%).

Like most trade organizations of its kind, the AAPS sends out a newsletter and occasional press releases. It publishes a journal (*The Journal of Paleontological Sciences*) and a periodic supplement (*The Paleontograph*).

The AAPS also sends out fundraising appeals to fund its public-education and, occasionally, lobbying efforts. In June 2020, *Fossil News* received one of these appeals, which also invited us to renew our membership.

This year, though, and for the foreseeable future, we're letting our membership lapse.

The reason isn't just that this latest communication from the AAPS is written in what has become the ambiguous and alarmist style of political fundraising appeals, whether you get them from the right or the left, from MoveOn or from the RNC: there is a terrible danger in the land, "they" are about to take something away from you, and you must send money if you want to "protect your rights." ("Our business and private collecting is under attack," the AAPS appeal reads.)

And the reason isn't that the AAPS has developed the habit of talking rather incessantly about the rights of fossil dealers but has almost nothing to say about their obligations or responsibilities to a wider community of people.

Yes, the AAPS slogan, "Supporting Paleontology through Ethical Commerce, Education, and Cooperation" is right there on the web page, but what AAPS's "Commercial Paleontology Code of Ethics" defines as ethical behavior for member merchants is both vague and substantially toothless. In other words, it doesn't address the real ethical problems in the fossil "industry" today nor does it seem intended to require fossil dealers to operate in a context of obligation to others who aren't fossil dealers—at least not very seriously. Here are the Code's ten points:

1. Strive to stay informed of and comply with International, National, State/Provincial and Local regulations pertaining to collecting activities and general business practices.
2. Obtain permission from landowners or governmental authorities to gain access to collecting sites.
3. Assure that all lands, properties, flora and fauna are left without damage to property or ecology as a result of the collecting activities.
4. Require that fossil materials received from outside collectors are obtained in compliance with the above collecting guidelines set forth by the Association.
5. Report to scientific experts any significant discoveries of scientific or public interest.

Fossils for sale at the annual Tucson Gem & Mineral Show. (Used via Creative Commons 3.0 license.)

6. Strive to place specimens of unique scientific interest into responsible hands for study, research and preservation.

7. Make no misrepresentation as to identity, locality, age, formation, repairs or restoration of paleontological specimens.

8. Conform to professional business practices when obtaining and disposing of specimens.

9. Maintain a good credit standing among fellow suppliers of earth science materials.

10. Encourage good relations and cooperation with agencies, institutions, and organizations actively involved in paleontological pursuits.[1]

---

[1] Compare the modifications to this version of the AAPS Code of Ethics with the version published along with the 1987 recommendations of the Committee on Guidelines for Paleontological Collecting of the National Academy of Sciences: "All members of the American Association of Paleontological Suppliers [as the AAPS was then called] will: 1. Stay informed of and comply with all Federal, State and Local regulations pertaining to collecting activities and general business practices; 2. Obtain permission from land owners or governmental authorities to gain access to collecting sites; 3. Obtain approval of Tribal as well as Federal authorities for access to lands within Indian reservation boundaries; 4. Assure that all lands, properties, flora and fauna are left without damage to property or ecology as a result of collecting activities; 5. Take every precaution to guard against fire and remove all litter from study or collecting areas; 6. Encourage the use of safety procedures and protective equipment in potentially hazardous collecting areas; 7. Require that all fossil materials received from outside collectors are obtained in compliance with the above collecting guidelines set forth by the Association; 8. Report to proper local authorities any significant discoveries of scientific or public interest; 9. Strive to place specimens of unique scientific interest into responsible hands for study, research, and preservation; 10. Make no misrepresentation as to identity, locality, age, formation, repairs or restorations of paleontological specimens; 11. Conform to professional business practices when obtaining and disposing of specimens; 12. Maintain a good credit standing among fellow suppliers of earth science materials; 13. Encourage good relations and cooperation with agencies, institutions, and organizations actively involved in paleontological pursuits" (54-55).

The mosasaur skull Russell Crowe bought from Leonardo DiCaprio in 2008 for between $30,000 and $35,000.

That's a lot of striving and encouraging, neither of which, of course, necessarily means *doing*. More importantly, though, the AAPS engages in no regular, active monitoring (as far as I know) of adherence to the Code of Ethics among its members, nor does it have an enforcement mechanism (other, one supposes, than canceling someone's membership).

Does the AAPS intend for its members to "require" anyone who sells or trades fossils to them to prove that she or he has complied with applicable law, has obtained permission to collect, and has done no damage to property or the environment in obtaining fossil material? If so, how? (Note, too, that damage to "property or ecology" isn't the only issue; economic and physical harm to human beings is another consequence of some fossil-collecting practices, and such harm is taking place right now in more than a few places in the world.)

Does this mean that fossil dealers will refuse to sell or buy fossils from Russia, China, Brazil, Morocco, Australia, Myanmar (which still supplies a great deal of the amber for sale in the world, despite serious human-rights issues; see Barrett & Johanson, 2020, and below) or any of the majority of the countries in the world that consider fossils part of their "national patrimony" and make their sale, and sometimes their private possession, illegal? Should AAPS member refuse to participate in any fossil fair or mineral show where

there might be fossils or minerals obtained without permission or whose collection violated the law or damaged "property or ecology"?

And who defines "significant discoveries of scientific or public interest" or "specimens of unique scientific interest"? The dealer who, by definition, has a conflict of interest? Ditto with reporting to scientific experts. How does AAPS know whether this has been done? What action does it take in the event of a violation?

Finally, what of the many other words and phrases in the Code of Ethics that are left undefined? Anyone who has tried to get a fossil dealer to disclose locality information knows such data are rarely forthcoming, and "make no representation" isn't the same as "provide full information without being asked." In response to a specific, precisely phrased inquiry, most dealers will disclose whether a specimen has been restored, reconstructed, recreated, composited, or repaired, but relatively rarely is such information *volunteered*. What are "professional business practices," meanwhile? And in which profession?

The AAPS does indicate that it is in agreement and compliance with the recommendations of the National Academy of Sciences' Committee on Guidelines for Paleontological Collecting (which Peter Larson, a founding member of AAPS and its president from 1979-1986, helped write). Those recommendations, however, were issued in 1987 (Committee on Guidelines for Paleontological Collecting, 1987): before Sue the *T. rex* went to auction and before the Field Museum paid $8.36 million for it; before Eric Prokopi was sent to jail for a massive black-market operation that smuggled millions of dollars worth of fossils out of Mongolia; before the intense legal and legislative battle over the ownership of the "Dueling Dinosaurs," discovered in Montana; before the return of the "Fighting Dinosaurs," found in the Gobi Desert, to a museum in Ulaanbaatar; before eBay; before "celebrity fossils" began going to big-name Hollywood or tech-world collectors like Nathan Myhrvold,[2] Nicholas Cage, Russell Crowe, Bill Gates, and Leonardo DiCaprio,[3] to corporations for

2 Myhrvold owns a *T. rex* skeleton that is displayed in the living room of his $31 million mansion in Medina, Washington.
3 "Actor Russell Crowe is auctioning off more than 200 of his personal items following his divorce from Danielle

display in their lobbies,[4] or to deep-pocket enthusiasts abroad,[5] all of whom commonly outbid museums for rare and unique specimens (see Helmore, 2019).

Yes, people have been picking up fossils for literally thousands of years, but more people are doing so all the time, and there can be no question that the increased visibility of the commercial fossil trade has fueled both interest in how much fossils are "worth" economically and the desire to own them. Though private fossil collections once largely housed fossils found by individual collectors, they can now be made up exclusively of fossils purchased online or at vast trade shows. They may come from nearly anywhere in the world and are frequently accompanied by only the vaguest of information about provenance. Fossils have become the new postage stamps.

As early as 2009, Matthew Carrano, curator of Dinosauria at the Smithsonian Museum of Natural History, told *Smithsonian* magazine:

In terms of digging for fossils, there are a lot more people than there used to be. Twenty years ago, if you ran into a private or commercial fossil prospector in the field, it was one person or a couple of people. Now, you go to good fossil locations in, say, Wyoming, and you find quarrying operations with maybe 20 people working, and doing a professional job of excavating fossils. (Webster, 2009)

Ten years later, Carrano was making the same point to BBC News:

With the rate of dinosaur discoveries showing no sign of slowing, [Carrano] says a private find may slip under the radar as "there's no catalogue or account of them aside from the occasional high-profile example that might make the news. The rest are sold and bought without any publicity and just disappear into private hands, often without science knowing they exist." (Timmins, 2019)

Elizabeth Jones, of the Department of Forestry and Environmental Resources, North Carolina State University, updated Carrano's observation further in 2020:

[T]he number of commercial companies driving the legal, and sometimes illegal, selling of fossils is estimated to have doubled since the 1980s. Not only is the commercialization of fossils increasingly prevalent ... some scientists have gone so far as to describe "the battle against heightened commercialization of fossils to be the greatest challenge to paleontology of the twenty-first century" .... To be clear, the commercial collection of fossils is not new. But it is different. Today, commercial fossil collection has changed as the social structure of the paleontological community has changed, giving rise to new motivations, interactions, and justifications of authority over fossil objects among different stakeholders both within the community and along its periphery. (Jones, 2020: 2-3)

It's partly because of these realities that the Guidelines for Paleontological Collecting simply don't reflect the current state of the fossil market,[6] the realities of increased private ownership of fossils, or the proliferation of commercial collectors. The Guidelines' authors noted that their "recommendations are designed to *reduce rather than promote regulation*" (Committee on Guidelines for Paleontological Collecting, 1987: 3, emphasis in original). That may have been an appropriate stance at the time, but it isn't today.

Among the issues that the AAPS Code does not acknowledge is that harm could ever come from collecting fossils for the purposes of selling them or from the commerce in fossils, and it insists, by implication, that commercial fossil collection is, at worst, neutral in its effects on the field of paleontology and, at best, a helpful and indispensable adjunct to science.

This last claim is an inescapable AAPS talking point. The *Journal of Paleontological Sciences*, founded by the AAPS in 2006, hosts a web page entitled "Fossil Specimens Placed in Museums and Universities by Commercial Paleontology." As of its June 2020 update, the page listed 145 separate donations by "commercial paleontology" of original fossil material to museums across the U.S. as well as in France, England, Scotland, Japan, Canada, Switzerland, Austria, Wales, South Korea, Ireland, The Netherlands, and Taiwan, noting that these gifts are "only the tip of the iceberg" (Association of Applied Paleontological Sciences, 2014/2020).

In an editorial to *Nature* in 2007, then-AAPS President Mike Triebold expounded on this claim:[7]

Spencer, including .... the mounted skull of a Mosasaur that Crowe 'acquired from' DiCaprio in 2008" (Lynch, 2018).

4 "In Dubai, an 80-foot-long *Diplodocus* is the star attraction of a shopping mall. And in Santa Barbara, California, one of the best *Tyrannosaurus* skulls ever found sits in the lobby of a software company, glowering, fangs bared, at the indifferent receptionist seated just opposite" (Conniff, 2019).

5 "There are no typical customers," says art dealer, Luca Cableri, "who sells dinosaur fossils among other curiosities through his gallery, Theatrum Mundi, in Arezzo, Italy, though "he has sold relics to several castle owners in France who were looking to spruce up their homes and attract more visitors" (Brown, 2019).

6 Between 2018 and 2019, eBay reported a 42% increase in the sale of mosasaur pieces in the UK and a 22% increase overall in fossil sales (Timmins, 2019). A conservative estimate of individual and small-business pages dedicated to the sale of fossils on Facebook is more than 290 (in English), not counting untold numbers of fossil-related groups whose members advertised fossils for sale.

7 Triebold was responding to an article in *Nature* by editor Mike Hopkin (2007) that had criticized the AAPS for,

By the 1870s, professional collectors were busy filling museums with dinosaurs and other fossils, by accepting the risks of exploration, discovery and excavation, then selling their discoveries, and in some cases collecting fossils on a contract basis. Visit any number of prestigious institutions and you will see magnificent displays whose very existence is owed to professional collectors. (Triebold, 2007)

Without contesting the main point—professional and contract collectors absolutely did fill U.S. and European museums with fossils in the second half of the nineteenth century and still do today—Triebold leaves out essential details.

In his book, *Dinosaurs and Indians: Paleontology Resource Dispossession from Sioux Lands*, Dr. Lawrence Bradley, professor of Geography/Geology at the University of Nebraska at Omaha, and arguably the country's foremost expert on the history and present of paleontological exploration and exploitation of indigenous lands, convincingly adduces evidence for the proposition that "fossils were yet another resource dispossessed from unsuspecting and impoverished Native Americans" (2014, 16).

Writing specifically in the context of Lakota (Sioux) lands in the decades after an 1825 treaty, Bradley argues that a large number of the specimens that built and enriched museums and helped create the careers and reputations of more than a few nineteenth- and twentieth-century paleontologists came from Sioux territory, beginning with the description of a "paleotherium" from an area east of the Black Hills (present day South Dakota) described in 1846 by a medical doctor, Hiram A. Prout (19). Prout

obtained the specimen from a friend at a trading post on the Missouri River. News of the find gave rise to a "fossil rush" that brought "a proliferation of scientists exploring throughout Sioux territory and reaping fossils, all without permission of the people who had sovereignty over the territory" (2014, 20). (See, in particular, Bradley's chapter entitled "Historical Geography of Fossil Dispossession from Sioux Lands, 1847-1899"; and Chew, 2005.)

"Commercial fossil trading in the United States started with quarrymen in New Jersey selling fossils to Joseph Leidy during the mid-1800s," Triebold wrote in *Nature*. True, but Leidy, the "Father of American paleontology," also had agents in the west. John Evans, who had been authorized and funded by the U.S. Congress to survey the "White Badlands" area; David Dale Owen, the American geologist who conducted the first geological surveys of Indiana, Kentucky, Arkansas, Wisconsin, Iowa and Minnesota; and Thaddeus Culbertson, who was sent to the upper Missouri region by the Smithsonian in 1850 where he collected ammonites, scaphites, fossil turtles, extinct rhinoceros, oreodonts, saber-toothed cats, and many fossil mammals from Sioux lands, a number of which ended up in Leidy's hands (Leidy specifically acknowledged the collectors Culbertson, Owen, and Prout in his writings; see Bradley, 2014: 10). By 1869, Leidy had named at least eighty-four species of fossil mammals from the Sioux "badlands" (Bradley, 2014: 26).[8]

---

in part, giving its new publication a misleading name. Because the *"Journal of Paleontological Sciences"* was a trade journal written by nonscientists for nonscientists, Hopkin noted fears (not entirely unreasonable) among academics that the publication would "give a sheen of scientific legitimacy to the dealings of commercial fossil hunters," and, less coherently, that it would energize the black market in fossils and encourage avocational and professional fossil collectors to keep important specimens out of the public domain.

[8] In the context of Triebold's and AAPS's insistence that commercial fossil dealers only help science, it's worth noting that Leidy, at the time director of the Academy of Natural Sciences, wrote in a letter in the late 1880s: "Professors [Othniel Charles] Marsh and [Edward Drinker], Cope, with their long purses, offer money for what used to come to me for nothing, and in that respect I cannot compete with them" (quoted in Milstein, 1992). Lazerwitz cites additional cases in which commercial interests preempted scientific research: A Harvard University dig in was disrupted when the site was raided, removing most of a dinosaur fossil that the academics had uncovered; a *Hadrosaurus* nesting site on a private ranch in Montana was lost to science when a commercial dealer offered the landowners more money for access to the site; the North

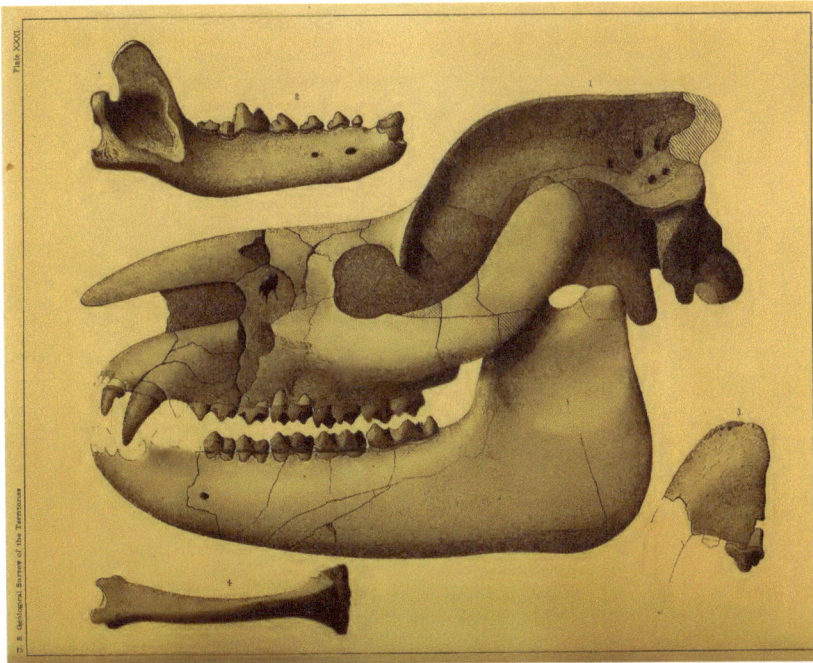

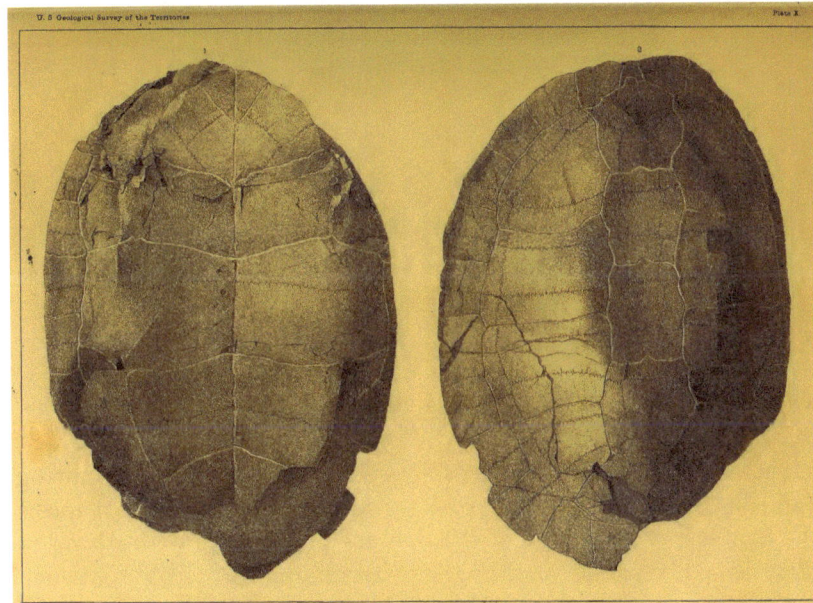

Some of the "Indian Territory" fossils studied by Joseph Leidy (from his 1873 *Contributions to the Extinct Vertebrate Fauna of the Western Territories*). Collectors in the "Mauvaises Terres" included Leidy himself as well as F.V. Hayden, James Van A. Carter, Col. John H. Knight, Dr. Joseph K. Corson, and "a Shoshone indian," who brought specimens to Carter. Top: Partial skull of *Palaeosyops*, a brontothere, along with assorted bones of other species. Bottom: Shell of an Eocene pond turtle, *Emys wyomingensis* (now called *Chrysemys*) collected near Ft. Bridger, WY, by Prof. Hayden in 1870. The Ft. Bridger area was within the forty-four-million-acre Eastern Shoshone homeland according to an 1863 treaty but, in 1868, over 70% of that land was taken back, and the Shoshone were restricted to the Wind River Reservation in east central Wyoming.

The American fur trader, Edwin Thompson Denig, who spent twenty-three years in the "Indian trade" in Sioux country (roughly from 1833 to 1856), wrote prolifically of his experiences with scientists, explorers, and naturalists, including John James Audubon, who came to the area in search of ornithological, paleontological, botanical, geological, ethnological, archaeological, and all manner of other specimens. Denig, whose work was not published until 1961, commented:

At the lower extremity of the Black Hills commences the Mauvaises Terres or Bad Lands.... The hills [of the Mauvaises Terres] are composed of clay and earths of different colors in strata running parallel through it lengthwise. It is in this part of the Sioux country the immense petrified turtles are found, some of which are estimated to weigh two thousand pounds. Many other petrifactions of fish, worms, and shells are also seen, most of which are said to have been submarine. Gigantic fossil remains of animals now extinct have also been discovered here by naturalists and traders. (1961, 6-8).

The tradition of collecting from indigenous lands continued as other treaties were written, broken, and renegotiated. Ferdinand V. Hayden, hired to "explore Sioux country before [the Indians] become enraged of encroaching white men" (Bradley, 2014: 25, quoting Lanham, 1973: 39), was a major collector, and Thaddeus Culbertson, James Hall, Edward Drinker Cope, Othniel Charles Marsh, and many others either took fossils from treaty lands, paid people to take fossils from treaty lands, or worked on fossils taken from treaty lands (see Bradley, 2014, 25-42 and passim). Marsh also desecrated Indian burial sites and brought human skulls back to the Peabody Museum.

It may be tempting to consider these late-eighteenth- and early-nineteenth-century collecting practices as "acceptable for the era," though they were, in fact, generally not acceptable to indigenous people, but it remains important to recognize that they established a precedent that has continued to the present era. Ex-

---

Dakota Geological Survey reported that a rancher in the state attempted to excavate a rare *Torosaurus* on his land, destroying it in the process, after he learned the potential value of the fossil (1994, at notes 33-35).

Ivory confiscated in the New York area and destined for destruction in Central Park in August 2017. Seen here is a small part of the two tons of tusks and carved objects that were pulverized in the "Ivory Crush," estimated to have represented around 100 elephants—or just over the number believed to be killed by poachers each day.

amples include the cases of Sue the *T. rex*, excavated on contested Cheyenne land; the *Dilophosaur* found on Navajo land in 1942; the *Titanotherium* fossils taken from confiscated Sioux territory in South Dakota in 2002 (see Mayor, 2007); and a plesiosaur collected on the Santee Sioux reservation in Nebraska in 2003 (Bradley, 2014, 177-181 and passim). Mayor noted that the sale of Sue in 1997 "spawned a soaring market for spectacular fossils among rich private collectors, leading to increased poaching from reservations and public lands" (2007: 8).

In 2007, the Nebraska Commission on Indian Affairs attempted to advance legislation covering non-human fossils found on Native American lands (fossils are not included in the 1990 law known as the Native American Graves Protection Act) (Dalton, 2007), though the effort apparently went no further than a proposition. In 2015, the Standing Rock Sioux was the first tribe to establish its own Paleontology Resource Code (Standing Rock Sioux Tribe, 2015).

Benjamin M. Eagle, fossil preparator for the Standing Rock Sioux Tribe, told *Prehistoric Road Trip* in 2020 that "There's been fossil looting here for decades. We hear about rumors of people going to South Dakota to sell fossils that belong to the tribe and even some pa-

leontologists coming to [tribal land] and taking fossils" (Graslie & Gimbel, 2020).[9]

All of that having been said, no one is disputing either the generosity and scientific interest of many commercial fossil dealers or their contributions to museum-building. The argument, instead, is that such acts do not tell the whole story of the commercial-fossil industry. Not only does commercial digging sometimes interfere directly with scientific research, but focusing solely on the "good actors" obscures the fact that the origins of the "commercial fossil trading" that Triebold cites lie in access to paleontological material from indigenous territory that was treated like many other Indian "goods" at the time: available for trade or sale or,

---

[9] Eagle's accusation that professional paleontologists sometimes disregard regulations regarding fossil collecting is not without basis in fact. A federal indictment in November 1993 "named the Smithsonian Institution in Washington and the Field Museum of Natural History in Chicago as buyers of fossils excavated illegally from Federal lands.... [The museums were not prosecuted] because they had cooperated with investigators" (Browne, 1994, C9). Given that the museums were "buyers" of fossils, the sellers were presumably commercial dealers.

at times, simply free for the taking. If that is the history of the practice of "commercial paleontology," it isn't necessarily a proud one.

THE IVORY TOWER

The AAPS is committed to ensuring that laws and regulations relevant to fossil collecting or sale aren't ill-conceived, unfair, or unduly harsh. For a trade organization, that's a more than reasonable goal. But it isn't hard to conclude, in reading through AAPS materials over recent years, that the organization's objective is to do away with *all* limitations on fossil collecting and selling. Or, in any case, you'll search in vain if you try to learn what the association believes might be fair, acceptable, and responsible limits on fossil commerce and for-profit collecting.

Right after the line about the attack on "business and private collecting," for example, the June 4, 2020 membership appeal segues into a discussion of an example of one of those "attacks": laws "passed to restrict the sale of many ivory items, including, in some cases, those made of fossil ivory, to stop the illegal trade in elephant ivory":

> The idea behind these laws is that smugglers are importing poached elephant ivory as fossil ivory such as mammoth tusk. Saving the elephants from extinction is a noble cause but many of these laws are being written so as to have much wider consequences.

Perhaps the point is a picky one, but the language in this brief paragraph is slippery. The Convention on International Trade in Endangered Species of Wild Fauna and Flora (CITES), with which U.S. Fish and Wildlife Service rules comply, does not regulate the trade in fossils or extinct animals. (CITES is an agreement among governments to ensure that international trade of endangered and protected wild animals and plants does not occur; it banned the blood ivory trade in 1989. The United States became a party to the Convention on January 14, 1974.)

Second, the language of the U.S. Endangered Species Act defines "antique ivory" as any ivory over 100 years old and specifically excludes it from prohibitions on the import, export, or sale of ivory, but does not mention mammoth or "fossil ivory." As a consequence, federal laws in the U.S., and, specifically the laws enforced by the Fish and Wildlife Service, also do not specifically "restrict the sale of fossil [ivory] items," meaning, presumably, mammoth or mastodon tusks, although "ivory" is a somewhat vague term that can include teeth, tusks, and bone from walrus, sperm and killer whales, narwhals, hippos, and even wart hogs, as well as critically endangered elephants).

Third, the AAPS statement implies that neither the "noble cause" of "saving the elephants from extinction" nor efforts to reduce smuggling and "ivory laundering" should be considered important enough to interfere with profit. That's a statement of values, not of fact, and there's something ethically questionable about arguing that the most important impact of ivory bans is that they make it hard for a comparatively few people to conduct legitimate business. The entire point of laws like the so-called "ivory ban" is to interfere with the incomes of bad actors and to discourage unwanted behavior through the use of economic and legal sanctions.

In any case, ivory import/export prohibitions are intended to affect the importation and sale of "blood ivory" *in the service of the goal of protecting endangered living species.* Groups including the NRA, the Knife Rights organization (Knife Rights, 2016), and some commercial fossil dealers, are outraged by this requirement, however, which they consider an unfair constraint on "enterprise."

One of the articles that AAPS lists as a resource on its "Fossil Ivory" page is a rant by the editor-in-chief of *Ammoland Shooting Sports News*, Fredy Riehl, regarding the "draconian law" that prohibits both mammoth and blood ivory and occasioned a seizure of ivory in New York from a "senior citizen jeweler" and a "folk art dealer." For anyone not inclined to read past the title, he gave his editorial an unambiguous one: "Government Bullies Nail First Victims" (Riehl, 2015).

Still, it is neither hypothetical nor secret that nefarious dealers label modern ivory as "fossil ivory" in attempts to get around the ban. Dr. Edgard Espinoza, Deputy Director of the National Fish and Wildlife Forensic Laboratory in Ashland, Oregon, confirms that dyeing or staining of illegal ivory is "widely practiced" in smuggling efforts, in part because of a common, though erroneous, belief that fossil ivory can be distinguished solely by color.[10]

In 2016, *National Geographic* reported several cases of illegal ivory being passed off as mammoth ivory, quoting Lucy Vigne, an ivory trade researcher and co-author of a 2014 report commissioned by the nonprofit group Save the Elephants, that "Sellers in Beijing also sometimes try to pass off mammoth ivory as elephant ivory in the store. They can pretend it's mammoth ivory if they're pushed" (Actman, 2016). *Popular Science* in 2019 noted that "evidence already exists that illegal elephant ivory is being intentionally mislabeled as legal mammoth ivory" and cited cases in China as well as the felony conviction of a New York-based antiques dealer who, in 2017, sold "elephant ivory mislabeled as 'carved mammoth tusks.'" Simon Nemtzov, a wildlife ecologist at the Israel Nature and Parks Authority and Israel's Scientific Authority for CITES, told *Popular Science*: "We know there is a trade in mammoth ivory [and] we know that some is being intentionally mislabeled [to hide] elephant ivory." (Nuwer, 2019).

Ample evidence exists, then, that mammoth ivory is being used to "launder" blood ivory (see Nuwer, 2019, e.g., as well as Larney, 2016, whose study in Hawai'i, one of the states that bans mammoth ivory, showed that "ivory from recently poached elephants is regularly smuggled

---

10 E. Espinoza, (personal communication, 6 July 2020).

into the United States and sold both in storefronts and online, often disguised as antique, legal ivory." An investigation by TRAFFIC, a wildlife trade-monitoring non-profit, "found that sellers in China, [Myanmar], Hong Kong and the US were mislabeling elephant ivory as mammoth ivory" (Kramer et al., 2017; Flanagan, 2019). Another investigation in 2018 "tested 109 'antique' ivory products legally purchased from countries all over Europe, including Germany, Ireland and Spain. Using

These facts alone might justify putting ivory sales under increased scrutiny, or even a ban. No, they are not direct proof that trade in mammoth ivory causes poaching of African elephants, but there is good reason to see a correlation. That's more than sufficient for a reasonable person to have doubts. And it's not the same as "not a shred of evidence."[12]

Perhaps, then, at least some of AAPS's efforts should be dedicated to unscrupulous traffickers and vendors who

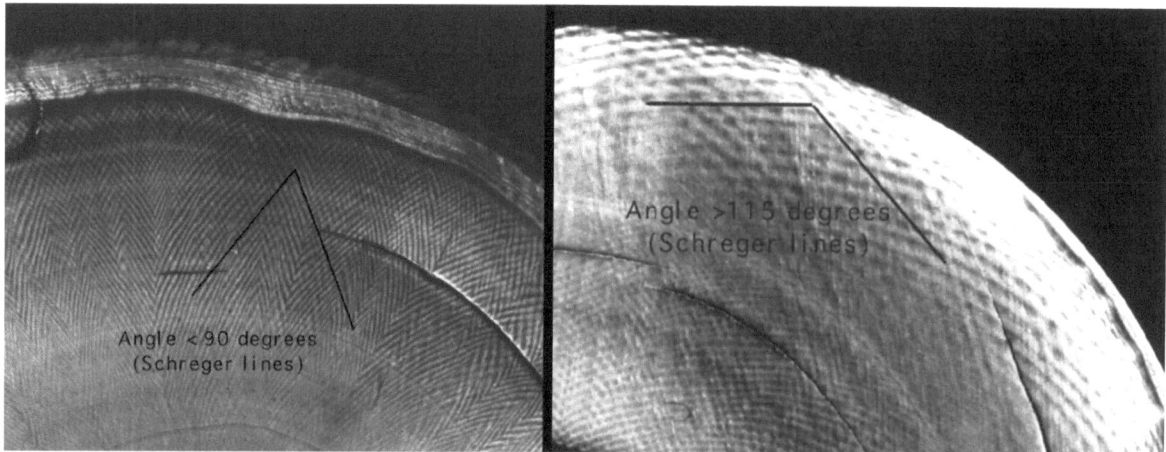

"Schreger lines"—chevrons or hatch marks—are one mechanism for distinguishing mammoth ivory from elephant ivory. On the left: a cross-section of mammoth ivory; at right extant elephant ivory. Ivory from other species can also be identified by examining cross-sections. From the website of the U.S. Fish and Wildlife Service's Forensic Laboratory.

radiocarbon dating, the team found that 75 percent of samples came from after 1947 [and were therefore illegal], some from animals alive in the 2000s" (Collins, 2019).

Though the AAPS source, Fredy Riehl, claims there is not a "shred" of evidence that bans on mammoth ivory have "saved a single elephant," the fact is: he simply doesn't know, nor does the AAPS or anyone else. Virtually no scientific research has been done on this specific point. What is known is that poaching of elephants either remains at previous levels or has increased in the majority of African countries. According to a 2020 study published in *Scientific Reports*, "For Africa as a whole, poaching did decline for 2011-2018, but the decline was entirely due to Eastern African sites. Our results suggest that poaching for ivory has not diminished across most of Africa since 2011" (Schlossberg et al., 2020). It's illogical to imagine that 15,000-20,000 elephants continue to be killed each year for some reason other than their ivory, and studies suggest that, when the worldwide price of ivory has increased, or demand has increased, so has the poaching of living elephants.[11]

have made things tough on virtuous dealers and to the development of strategies for active involvement in curtailing smuggling, laundering, and other illegal activities.

ICE IVORY

The National Fish and Wildlife Forensic Laboratory's Edgard Espinoza explains that wildlife inspectors are deployed at major ports of entry across the nation and are well trained to distinguish extinct from extant species. If a consignment is believed to be illegal, they detain the material and send it to the U.S. Fish and Wildlife Service Forensics Laboratory for analysis. If the material turns out to be legal, it is returned to the owner. If it is not, various scenarios can play out, including "abandonment" of the shipment in question. Such determinations are typically made quickly, though they sometimes do take months, Espinoza acknowledges, if the lab is particularly busy.

---

[11] "[T]he ivory trade (and therefore elephant poaching activity) is driven by demand.... Higher ivory market prices lead to higher poaching incentives, and therefore greater numbers of elephants being killed" (Sosnowskia, 2019:392); "[According to] Robert Hormats, a senior State

Department official, 'Without the demand from China, this [poaching increases in elephant-rich areas] would all but dry up'" (Gettleman, 2012); "A Zambian wildlife crimes tracker told *Oxpecker*s that poachers who have flooded the region over the past three years are responding directly to heightened demand for ivory from locally based Chinese ivory trafficking syndicates" (Nkala, 2016).

[12] Evidence may be coming, however. CITES failed to adopt a total ban on mammoth ivory in 2019 but will reconsider the issue in 2022, after a study of the effect of the mammoth ivory trade on global ivory markets has been completed.

Claims are occasionally made that the ivory ban is unfair because the burden of proving that material is exempt from the ban falls on the possessor of the questioned ivory, but the facts are somewhat different. If the possessor claims the ivory is "antique" but not fossil, then, yes, evidence of the provenance of the item is required, which can include family photographs, letters, or any other documentation that shows that the ivory in question was legally harvested one hundred or more years ago. If the question involves distinguishing fossil from extant material, the U.S. Fish and Wildlife Service Forensics Laboratory does the testing at no charge. According to Espinoza, however, no case has occurred since the ban was enacted in which legitimate mammoth ivory was detained, tested, and then mistakenly certified as illegal modern ivory. That's an important point.

Given that fossil dealers are typically not dealing with "antique" ivory but with much older fossil or subfossil material, in any case, the federal requirement barely concerns them. Even if it did, however, what is the nature of the unique and excessive burden that would be placed on their businesses if they were asked to demonstrate the provenance of ivory they bought, sold, or traded? We must carry proof of insurance in our cars when we drive. Someone who adds a story to a house needs a permit; an inspector can stop work or levy fines if the permit isn't available. No one simply takes air travelers at their word if they say they aren't bringing weapons onto a plane; they must demonstrate the fact. Proof of American citizenship must be shown by anyone seeking a U.S. passport; no one is required to trust an applicant's claim to be a citizen without documentation.

The AAPS is nonetheless absolutely correct when it claims, as it does on its website, that some states (eight, according to the AAPS's online list) have passed laws that specifically *include* "mammoth ivory" in prohibitions. (Other states have considered doing so, but most of the draft laws that name "mammoth ivory" have languished in legislative committees for years.) This is the boilerplate paragraph that a number of U.S. jurisdictions have copied into adopted or proposed legislation:

"Ivory" means any tooth or tusk composed of ivory from any animal, including, but not limited to, an elephant, hippopotamus, *mammoth*, narwhal, walrus, or whale, or any piece thereof, whether raw ivory or worked ivory, or made into, or part of, an ivory product" (emphasis added).

Most states' laws, however, specifically exempt "antique" ivory (i.e., over 100 years old), which clearly includes fossil ivory. Where that is not the case, Espinoza blames ignorance—specifically, the misconception that mammoth ivory cannot easily be distinguished from elephant ivory. As early as 1990, in fact, Espinoza co-pioneered a laboratory procedure to differentiate mammoth

and mastodon ivories from modern elephant ivory, and his forensics lab won an international award for best research paper when he and his research partner presented their method at the International Association of Forensic Sciences meeting that year. The full report was published in the *Journal of the American Institute for Conservation* a few years later (Espinoza & Mann, 1993). Espinoza and Mann also prepared the extremely detailed *Identification Guide for Ivory and Ivory Substitutes*—in use for the last ten years by the U.S. Fish and Wildlife Service Forensics Laboratory. The guide was "developed to give information about a nondestructive and visual means of tentatively distinguishing clearly legal ivory from suspected illegal ivory at ports of entry," which approach is not meant to take the place of "an examination of the ivory object by a trained scientist ... to [positively identify] the species source" (Espinoza & Mann, 2010).

If it is not difficult to tell legal mammoth ivory from illegal "blood" ivory, then, there's no real reason beyond hyperbole to stoke fears that mammoth ivory will be erroneously confiscated and, arguably, no reason for blanket state bans on mammoth ivory either. In fact, in October 2017, Republican Senators Dan Sullivan and Lisa Murkowski of Alaska introduced the so-called "Allowing Alaska IVORY Act," which, among other provisions, would have prohibited individual states from issuing bans on mammoth-ivory products that were produced by Alaskan natives as "an authentic native article of handicrafts or clothing" ("Senate Report," 2018). As of this writing, the bill has remained on the legislative calendar without further action since December 2018.

The vast majority of mammoth ivory for sale in the world does not come from indigenous Alaskan craftspeople, however, but rather from large mining operations in Siberia. As Doug Meigs reported in 2016:

Most of the world's untouched mammoth ivory remains locked in the frozen permafrost of Siberia. When snows melt during the brief Arctic summer (from mid-July to mid-September), riverbanks often reveal prehistoric remains. Warmer summers means the permafrost is thawed longer every year. That means more and more mammoth tusks are protruding from the ground.... Indigenous locals, seasonal tusk hunters, and Russian gangs aggregate the raw tusks in Siberia. Officially, the tusks must be approved for export by the government authorities, but traders (and smugglers) are increasingly taking their purchases directly into mainland China over the land border with Russia, Mongolia, or neighboring countries.

As much as 90% of this Siberian production, sometimes called "ice ivory," ends up in China, "which is also the main destination for illegal elephant ivory" (Actman, 2016). Zara Bending of the Centre for Environmental Law at Macquarie University in Sydney, Australia, noted that "[i]mports [of mammoth ivory] to Hong Kong have increased dramatically from fewer than 9 tonnes per year from 2000 to 2003 to an average of 31 tonnes per year from 2007 to 2013. Similarly, one survey found a fourfold increase in mammoth ivory sales in Macau [the "Special Administrative Region" of the People's Republic of China] between 2004 and 2015" (Bending, 2019).

Siberian mining operations by tusk hunters, or "tuskers," which are frequently illegal, improvised, and dangerous, can be found throughout the one-million-square-mile Yakutia region in Eastern Russia, but especially alongside waterways. According to Amos Chapple, who spent part of a summer "under cover" with tuskers in 2016, money is what lures bands of impromptu "commercial fossil dealers" to ransack paleontological sites: In an area "where the average salary is less than $500 a month," tuskers can earn $100,000 in eight days from the mysterious "agents" who roam the region's towns looking for mammoth ivory to buy.

Constantly on the run from police and environmental-protection officers, Chapple reported, tuskers rig up gasoline-powered high-pressure hoses designed for fire-fighting to pump water from waterways. With them, they blast vast caves and enormous tunnels into hillsides to expose fossils, leaving behind a ravaged landscape and filling bodies of water with hundreds of tons of silt, which has destroyed fishing and ruined Yakutia's rivers and lakes. Chapple added, "In status-mad China, mammoth ivory appears to be subject to an economic phenomenon whereby high prices drive increased demand, thus further rais-

ing prices. Once sculpted by a master carver [a pair of mammoth tusks] regularly sell for more than $1 million" (Chapple, 2016).

Quoted in *National Geographic*, Daniel Fisher, a paleontologist at the University of Michigan who studies woolly mammoths at dig sites in Siberia, confirms the impact on scientists: "Over the years, I've seen what I can collect dwindle quite significantly. There's still im-

"Tuskers" at work in the Yakutia Region. Above: To expose fossils, an entire hillside has been gouged out with high-pressure jets of water. Below: A mammoth hunter prepares to transport a tusk. Photos © Amos Chapple from his 2016 investigation, "The Mammoth Pirates," for RadioFree Europe.

portant questions to be solved about woolly mammoths. Do we study the tusks and learn something from them, or do we carve them?" (Actman, 2016).

Once again, however, the AAPS position makes it nearly impossible to grasp the equities fully or to balance interests. If it is true that the mammoth-ivory trade fuels the demand for elephant and other ivory in general;[13] if most of the world's prehistoric ivory comes from the Siberian permafrost and is frequently smuggled out or excavated under conditions that destroy paleontological or archaeological context and ruin the local environment; if it depends upon black-market trade; and, as many experts have claimed, if it is true that prehistoric ivory trade has become a "'cynical laundering mechanism for freshly poached elephant ivory'" (Meigs, 2016), perhaps bans that include mammoth ivory exist for other, defensible reasons—and not solely because of an irrational fear that fossil and illegal ivory cannot be distinguished from one another.

Israel and Kenya, for example, have asked the Convention on International Trade in Endangered Species of Wild Fauna and Flora (CITES) to regulate the trade in mammoth ivory (not to block or prohibit it) precisely because "elephant ivory is being sold as mammoth ivory [allowing] traders [to] skirt supervision [and] posing a serious threat to elephants" and because the trade in mammoth ivory "buoys smugglers of elephant ivory" and "has severely damaged the environment as would-be traders dig for skeletons [in Siberia]" (Rinat, 2018). The hope is that such supervision would "make sure intentional mislabeling doesn't happen so that the mammoth ivory trade doesn't contribute to laundering illicit ivory" (Actman, 2016) because, as Israel stated in its proposal, "the legal, unregulated trade in mammoth ivory facilitates illegal trade in elephant ivory" (Kukreti, 2019.

Here is an opportunity for AAPS to lead by joining conservation efforts, endorsing the Kenya/Israel proposal, opposing black-market fossil dealing, supporting a requirement that dealers display provenance documents, and promoting a reasonable standard for the industry rather than simply railing against laws intended to eliminate poaching and the illegal trade of animal parts.[14] But some fossil dealers seem to be having trouble understanding a simple truth: ivory bans aren't about them.

FOREVER AMBER

For decades, but especially in recent years, spectacular fossils found in Burmese amber—including dinosaur tails, birds, small skulls, ammonites, intact lizards, swarms of spiders and insects, and other vertebrate and invertebrate wonders—have excited scientists and the general public alike. A fact that often seems to be a footnote, if it is mentioned at all, is that these fossils are paid for with human suffering and loss of life.

Though amber is found in many places around the world, the deposits most commonly tapped for jewelry and science are in the Dominican Republic, the Baltic region,[15]

---

13 In 2019, the Animal Law Committee of the New York City Bar Association issued a formal *Report on How Mammoth Ivory Contributes to Elephant Poaching*, noting that "prohibiting trade in mammoth ivory makes it more difficult for traders of illegal elephant ivory to escape detection.... [T]o avoid being caught, smugglers often try to pass off elephant ivory as mammoth ivory or identify elephant ivory as mammoth ivory on shipment documentation.... Second, allowing trade in any ivory likely increases the overall demand for ivory, and poaching in turn rises to meet that demand.... Significantly, to the extent that recent state laws reduce the illegal ivory trade, they also help combat criminality and terrorism. The illegal ivory trade is closely linked to the drug trade, money-laundering, weapons trafficking, trading, and governmental corruption" ("Report on How Mammoth Ivory Contributes," 2019).

14 In fact, one could be forgiven for concluding that the AAPS sometimes seems to take a fairly untroubled view of black markets, which might be considered alarming for an organization that considers "ethical commerce" vital. In a 2014 article in *The Journal of Paleontological Sciences*, AAPS leaders and members Neal Larson, Walter Stein, Michael Triebold, and George Winters commented that "In Brazil, China and many places elsewhere in the world, fossils are not allowed to be collected without fines, imprisonment or worse. This has not stopped the local population from excavating fossils. In many cases these laws have simply driven fossil trade underground to the black market where no one benefits" (Larson et al., 2014). Of course, it's naïve to claim that "no one benefits" from the black market; in fact, someone always does—illegally. The Larson group was likely basing its opinion in part on the views of British paleobiologist David Martill, who, writing in 2011 about the smuggling of Brazilian fossils, said: "Sadly, Brazil as a country suffers from endemic corruption, which is a way of life in parts of the country.... This corruption permits a global fossil-exporting industry worth millions to flourish. Some people ... think such a trade is 'a bad thing' and want it stopped. I don't: I want to see it expanded.... [T]o protect Brazil's fossil heritage from commercial dealers ... you need to stamp out Brazil's grass-root corruption and that will never happen" (Martill, 2011). To make it plain, Martill suggests that the response to black-market fossil trading is not to institute the rule of law, prosecute corruption, lift people out of poverty, or retrain workers so they have options other than illegal commerce, but rather to abolish all laws designed to protect paleontological resources.

15 So-called "Baltic amber" often means amber from Russia and the Ukraine, where it is mined illegally and dangerously under conditions of gang warfare, organized crime, and widespread destruction of the environment: "[Amber-mining] gangs use water pumps ... to take advantage of amber's low density. In popular regions, hundreds of gangs may mine in the same area; thousands of men pump ... the soil full of water in the hope that amber will float to the top.... Illegal amber mining is causing untold environmental damage, leaving lands as barren as moonscape. Vast areas of forest have been stripped for the purpose, but it's

and Myanmar (formerly Burma). Of these, Dominican amber is the youngest, and Burmese amber is the oldest (perhaps 100 million years old, though the fact that amber mines are located in a conflict zone means that scientists have not been able to study and date the exposures accurately).[16] Burmese amber also has the distinction of including bird and dinosaur feathers, bones, small vertebrates, and other paleontological treasures that are uncommon or absent in amber from other areas.

But the harsh and shocking realities of the Burmese amber market—human-rights abuses, guerrilla warfare, smuggling operations, child labor, environmental destruction, genocide, mining deaths and injuries—are far from unknown.

According to Joshua Sokol, writing in *Science* magazine in 2019, the fossils in question

come from conflict-ridden Kachin state in Myanmar.... In Kachin, rival political factions compete for the profit yielded by amber and other natural resources. 'These commodities are fueling the conflict,' says Paul Donowitz, the Washington, D.C.–based campaign leader for Myanmar at Global Witness, a nongovernmental organization. 'They are providing revenue for arms and conflict actors,

Above: A piece of Burmese amber preserves microscopic filaments from the tail feathers of a dinosaur, one of the first specimens to allow a look at the feathers' actual structure. © Royal Saskatchewan Museum/ R.C. McKellar. *Facing page:* Amber prospectors in Kachin State, Myanmar. © *The Irrawaddy.*

and the government is launching attacks and killing people and committing human rights abuses to cut off those resources.' Much of the amber is smuggled into China in a trade that Tengchong officials and traders ballparked at between $725 million and $1 billion in 2015 alone.

Sokol added:

The amber mines are ... a cesspit of human rights and environmental abuses. 'Poor working conditions are pretty much a trademark in all the mines in Kachin state and environmental regulations, where they exist, are largely ignored,' says Hannah Hindstrom (a senior campaigner and Myanmar resources expert at NGO Global Witness). Amber is only part of the resource war, and scientists are far from the only people buying it. But it is impossible not to conclude that they are complicit, if not actively involved, in a trade that helps to fund a war.

According to *The Irriwaday*, a major independent news source for Myanmar and Southeast Asia, the Myanmar armed forces, the Tatmadaw, launched an offensive in the amber-mining region in 2017:

After a dinosaur tail which is believed to be 99 million years old was found in a piece of amber discovered in Tanai, the Tatmadaw launched attacks

---

the mining technique itself that completely decimates the soil, making it incapable of sustaining plant life.... Due to the unregulated nature of the mining, there are little if any safety precautions taken. The pits dug are often unstable and liable to collapse, crushing or drowning miners.... [T]here are widespread reports of corruption, both in the police forces and the government. Some miners claim that police officers are happy to turn a blind-eye, which may cost as much as 30% of your profit. And if you fail to pay, the police will raid your home and arrest you for illegal mining. This ensures that people become part of the 'protection' racket run by the police" (Lempriere, 2017).

[16] Export of raw amber from the Dominican Republic has been illegal since 1979 and, since 1989, amber with insect inclusions can't "be exported without the consent of the National Museum of Natural History [of the Dominican Republic]" according to Katherine Gammon's "The Human Cost of Amber" (2019).

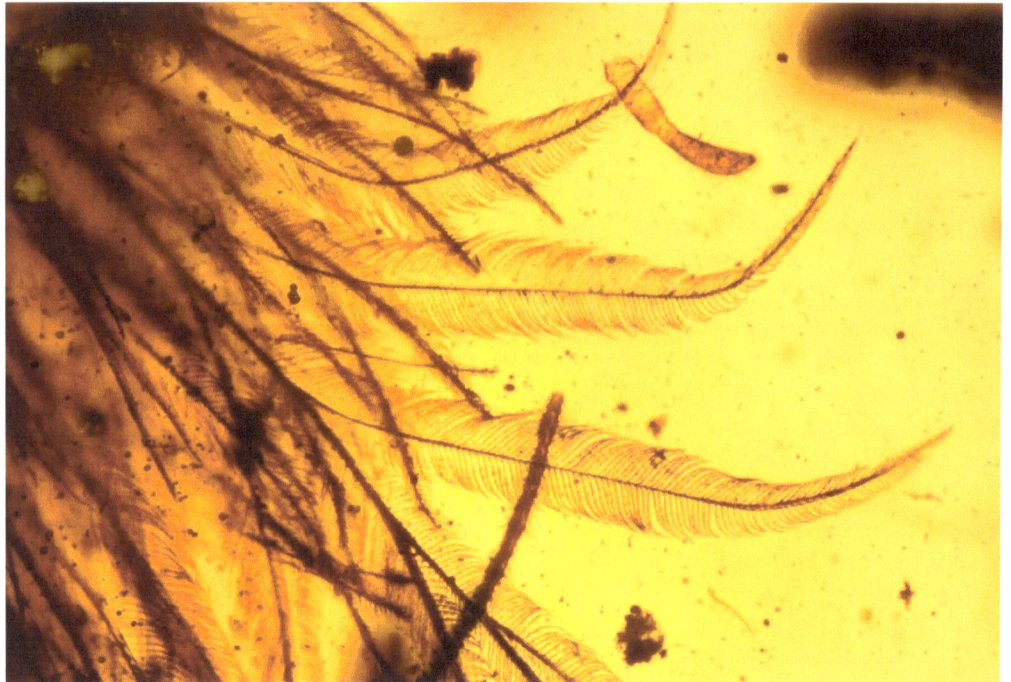

... on the pretext of concerns for environmental conservation and the risks of losing state revenue. The Tatmadaw dropped leaflets from airplanes, asking locals along with amber miners to leave the area.... Following the clashes, the Tatmadaw [took] control of the area.... Amber miners run the risk of being shot by soldiers who are watching from outposts built on hills, and stepping on mines planted by both sides.... Though it appears to people outside Tanai that the mines are closed, that is not the case, said local residents. On the ground, the military has built outposts and blocked all the roads leading to the amber and gold mines. But inside the mines, some businessmen who have bribed the military are still operating, locals said. ("Displaced by Clashes," 2019)

The Burmese amber trade has been the subject of articles and editorials in the *New York Times; The Atlantic; The New Scientist; Time; Science; The Financial Post;* the blog of well-known University of Portsmouth (UK) paleontologist and paleoartist, Mark Witton; and even the London-based trade journal, *Mining Technology,* among other outlets. More than a few scientists, in fact, including David Grimaldi, the curator of amber specimens at the American Museum of Natural History in New York; Brian Brown, the entomology curator at the L.A. County Museum of Natural History; Steve Brusatte, a vertebrate paleontologist at the University of Edinburgh

and author of numerous books, including *The Rise and Fall of the Dinosaurs*; and the co-Editors in Chief of the *Journal of Systematic Palaeontology,* have made clear that, for ethical reasons, they will no longer support or be involved in activity that promotes amber traffic from Myanmar; Grimaldi went so far as to write a brief letter to *New Scientist* in June 2019 in which he said "If amber sales are funding bloodshed [in Myanmar], then a strict boycott of Burmese amber is absolutely necessary."

In April 2020, the Society for Vertebrate Paleontology announced on its website that it "strongly discourage[d] its members from working on amber collected in or exported from Myanmar since June 2017" (Society for Vertebrate Paleontology, 2020), and, that same month, its Executive Committee sent a letter to the editors of some 300 scientific journals that is worth quoting at length:

One particularly alarming example [of fossils in and from conflict zones] is the so-called 'Burmese amber' that contains exquisitely well-preserved fossils trapped in 100-million-year-old (Cretaceous) tree sap from Myanmar.... Where Burmese amber is mined in hazardous conditions, smuggled out of the country, and sold as gemstones, the most disheartening issue is that the recent surge of exciting scientific discoveries, particularly involving vertebrate fossils, has in part fueled the commercial trading of amber. The rarest types of fossils are sought after for exceptionally high prices.... Our understanding is that the Myanmar military has recently seized control of the mining operation [and the offensive] has been condemned by the UN as a genocide and crime against humanity.... SVP regards the problem surrounding Burmese amber to be particularly pressing [and] we do not condone promoting our scientific endeavor at the expense of people facing humanitarian crisis.... In the case of Burmese amber, boycotting its commercial trading altogether, at least until the situation in the country stabilizes, may ultimately be one of the most effective solutions. Journal editors, publishers, and peer referees can also serve as 'gatekeepers' to set high ethical standards in our scientific field by only publishing manuscripts on Burmese amber that [was] acquired prior to the recent conflict. We also hope that scientists who study Burmese amber as well as private fossil collectors would exhibit the highest possible level of integrity so as not to encourage a black-market for commercial trading. (Rayfield, Theodor & Polly, 2020)[17]

Even Chinese celebrity scientist Lida Xing, who has published widely on fossils found in Burmese amber (he told Sokol that he "spends roughly $750,000 on Burmese amber per year") has acknowledged that the conditions at the mines are "inhuman." In 2014, he had himself smuggled, in disguise, into the Hukawng Valley to view an amber mine in operation:

> According to Xing's account, the 'mine' turned out to be a shanty town of around 3000 tents, each covering a narrow mineshaft up to 10 metres deep. The miners were living in bamboo huts among the shafts. Xing described conditions in the mines as 'very dangerous, inhuman.' The miners work with no safety equipment and often die of suffocation or in collapses. (Lawton, 2019)[18]

But that hasn't stopped Xing—or others—from continuing to buy amber from Myanmar and otherwise participating in the amber trade. And the justifications for such complicity will be familiar to anyone who knows what so often happens when human-rights considerations clash with the chance for career advancement or profit: The amber will be sold even if scientists don't buy

it (Xing: "If we don't get a specimen, it probably becomes cheap jewelry around some young girl's neck"; Sokol, 2019). Other important benefits exist beyond scientific discovery (Jingmai O'Connor of the Chinese Academy of Sciences: "These are the kinds of things that make people excited about science, that make children want to grow up to be scientists"; Joel, 2020). Taking a stand against the atrocities justified by the Burmese amber trade is someone else's responsibility (Anonymous online commenter: "Fossil amber accounts for only 1% of the demand and would not on its own be creating these conflicts.[19] The jewelry industry should be the first to take a stand").

In other words, it isn't that the sins of the Burmese amber market aren't known; it's that they're widely ignored—and even more widely justified by a sort of "fine people on both sides" logic: "I've weighed the ethical questions, and even if my participation in the Burmese amber market takes advantage of a conflict area where no rule of law protects citizens or workers, even if it fuels war and genocide and promotes child labor, and even if I am participating in a black-market Chinese smuggling operation, getting my name on scientific papers or making a profit is the greater good." That decision is a moral one.

As paleontologist Steve Brusatte commented to the *New York Times*, "We scientists need to make our own decisions based on our own ethics and values, and we could use guidance from our professional societies" (Joel, 2020). Some societies have, in fact, provided that guidance, and one organization that could serve such a function among non-scientists is the AAPS, the largest trade group of fossil dealers in the United States. Growing awareness of the human cost of the Burmese amber trade make this a perfect time for the organization to insist that its members do not buy, sell, or trade Burmese amber; do not participate in fairs or businesses where such commerce takes place; and publicly name dealers who refuse to comply.

SHOWDOWN AT THE "DUELING DINOSAURS" CORRAL

Virtually all of what is considered "private land" in the U.S. was originally occupied by long-time residents who predated Europeans (and the United States) by something like fifteen millennia. The land was bought, stolen, taken by force, or extorted from these people, and it was

---

[17] An electronic database search turned up fifty journals that published paleontological research based on Burmese amber between 2016 and the present. The majority are U.S.-based, but publications from China, the Czech Republic, Poland, Latvia, Germany, France, the Slovak Republic, Italy, Ukraine, and the UK are also represented. The fifty journals published 163 separate articles. The Elsevier journal, *Cretaceous Research*, whose editor-in-chief is Eduardo Koutsoukos of the Geology-Paleontology Institute for Geosciences at Heidelberg University, Germany, accounted for 76% of these. Without querying individual authors, it isn't possible to know whether published research was based on amber obtained since the explosion of commerce at the Tengchong bazaar or on historical collections of amber in museum and university repositories.

[18] Reporting the deaths of 126 people in a landslide at a jade mine in Kachin State, Myanmar, in July 2020, science writer Joe Bauwens noted: "Myanmar is the world's largest producer of jade, though much of this is produced (along with other precious and semi-precious minerals such as amber) at unregulated (and often illegal) artisanal mines in the north of the country, from where it is smuggled into neighbouring China. Accidents at such mines are extremely common, due to the more-or-less total absence of any safety precautions at the site. At many sites this is made worse by the unregulated use of explosives to break up rocks, often leading to the weakening of rock faces, which can then collapse without warning. The majority of people in this industry are migrant workers from the surrounding countryside, not registered with any local authority." (Bauwens, 2020).

[19] Various sources estimate the value of the Burmese amber market at $1 billion USD per year. The 1% figure is widely used to argue that the market for fossil-bearing amber is small. In fact, the statistic is misleading. Amber experts have noted that fossil inclusions are found in only about 1% of amber. That's not the same as saying that the sale of fossil amber accounts for only 1% of $1 billion (even if it did, $10,000,000 would be a significant figure.) In any case, the actual percentage of Burmese amber sales attributable to the fossil market is not known and probably cannot be.

next sold directly into private hands, granted to a state at statehood, or given away to individuals via the various federal Homestead Acts of the second half of the 1800s (about 10% of the total area of the United States was given away for free during that time; most of this previously public land was located west of the Mississippi River). Often, the "surface rights" to this land (in very general terms, what was on the surface of the land) and the "mineral rights" (again, very generally, what existed beneath the surface) were "severed," meaning that the two categories of rights could be sold or deeded separately. In that case, the owner of the surface rights did not necessarily own the mineral rights and vice-versa. This practice was particularly important in the American West, where there were known to be reserves of coal, petroleum, and other "minerals," because it protected the government's access to these resources on into the future.

rights) to the Murrays in 2005. Each of Severson's sons ran an extraction company interested in exploiting the coal, oil, and gas that lay beneath the ground. The Murrays, when they acquired the land in question, received one-third of the mineral rights and each of the two Severson brothers held one-third each.

In 2013, the Murrays attempted to sell the "Dueling Dinosaurs" at auction, setting their minimum price at six million dollars.[20] The auction, organized by the Bonhams auction house in Manhattan, was dedicated solely to fossils and, according to *Science* magazine, "Nearly 20% of the specimens, ranging from a large Eocene turtle from Wyoming to an Oligocene false saber-toothed cat from South Dakota, were rare and carried auction estimates between $100,000 and $2 million" (Pringle, 2014).

The Murrays' partners in the mineral rights, the Severson brothers, objected, claiming that their two-thirds share of the mineral rights entitled them to a portion of the proceeds from sales of fossils found on the land.

The original case was heard in the U.S. District Court for Montana and, in 2016, the court issued a finding that the fossils belonged solely to the

The "Dueling Dinosaurs" at auction. © Seth Wenig/Science. *Inset*: Closeup of vertebrae in the ceratopsian's tail. © Robert Clark/Smithsonian.

Murrays and were not included in the mineral rights. The Seversons appealed to the 9th Circuit Court of Appeals (which has jurisdiction over Alaska, Arizona, California, Hawaii, Idaho, Montana, Nevada, Oregon, and Washington), whose judges reversed the lower Court. The 9th Circuit was asked to reconsider its decision. This rarely happens, but, in this instance, the 9th Circuit agreed and sent the case to the Montana Supreme Court,

In 2016, these rather dry legal distinctions became the subject of a lawsuit over the ownership of (and right to sell) an exceptional specimen discovered in Montana: the "Dueling Dinosaurs," a twenty-two-foot-long tyrannosaur and a twenty-eight-foot-long ceratopsian that had apparently died in mutual combat.

The details that gave rise to the nearly five years of litigation that ensued have been widely reported and repeated, but here's a summary. In 2006, the "Dueling Dinosaurs" were found by Clayton Phipps and his cousin, Chad O'Connor, on ranch land whose surface rights belonged to Mary Ann and Lige Murray but whose mineral rights were shared with the sons of George Severson, the landowner who had sold the property (including surface

---

[20] The Murrays sold a nearly complete *Tyrannosaurus rex* found on the same land to a museum for several million dollars in 2014 as well as "a *Triceratops* foot for $20,000 and [they] offered to sell [a *Triceratops*] skull for between $200,000 and $250,000" (reported in *Murray and Murray v. BEJ Minerals, LLC, and RTWF, LLC*, 2020).

asking them to determine, once and for all, whether fossils belonged to surface or mineral rights as the State of Montana defined them. The Montana Supreme Court ruled they were part of the surface estate.

In the meantime, the Montana State Legislature had unanimously passed legislation declaring that fossils did not fall within the reservation of mineral rights in Montana. As a result of the Montana Supreme Court ruling, the 9th Circuit vacated its original opinion and concurred that the "Dueling Dinosaurs" were part of the surface estate of the Murrays property.[21]

As soon as the original 9th Circuit opinion was issued in late 2018, the AAPS cranked up its public-alert system to raise funds to file a "friend-of-the-court" brief

"That's crazy; of course fossils aren't minerals." The second was designed to elicit a no-less-automatic "don't tread on me" reaction from fossil dealers and collectors in Western states, many of whom are aware of the long history of tension between private landowners and government agencies over questions of land rights, access, and ownership. And the last was the tried-and-true fundraising tactic of the exaggerated emergency; like most of those, it was not precisely true. Still, fossil clubs around the U.S. dutifully forwarded the AAPS's calls-to-action in emails and newsletters to their own members.

But the original question was never whether fossils were literally minerals. What the 9th Circuit court held

A re-creation of what the Dueling Dinosaurs might look like once fully exposed. © Samuel Farar.

and to rally its members, and it did so with a degree of sensationalism that might have been good PR but got in the way of reasonable attempts to discuss the merits and potential repercussions of the case.

There were three main battle cries: "fossils are not minerals," "protect the rights of landowners," and "all fossils are in danger." The first counted upon a kneejerk response on the part of a public distrustful of nuance:

---

[21] "Dueling Dinosaurs" does not belong solely to the Murrays, however, because they contracted for a share of eventual profits with the discoverer and one of the excavators of the specimen, Clayton Phipps (who operates Dueling Dinosaurs LLC). Phipps, in turn, hired Peter Larson of the Black Hills Institute of Geological Research and the owners of CK Preparations, Chris Morrow and Katie Busch, to dig and prep the specimen. All of them will receive a portion of Phipps' share. See Pantuso (2019).

was that, in split-estate cases, fossils belonged to the owner of the mineral rights rather than to the owner of the surface rights. The distinction between "fossils are minerals" and "fossils belong to the holder of the mineral rights" may seem subtle, but, as a rhetorical device, the difference is enormous. The second makes a boring headline; the first fans natural skepticism and makes fact and detail secondary.

As far as the question of protecting landowners, the American reverence for "private property" deserves some thoughtful exploration. So, likewise, do the reasons for which the "right" of property ownership is so often considered more important than other rights—including those of the public to have access to important scientific information. In fact, the American legal system already treats property rights as more important and accords them more protection, on balance, than it does many other kinds of human rights.

Perhaps this is rooted in three founding notions of the United States: that only white, land-owning men were eligible to vote or to participate in politics; that property could simply be taken from the people who occupied it before Euro-Americans wanted to live on it, farm it, log it, or mine it; and that some men, women, and children *were* property. It's also worth noting that, in the Declaration of Independence, Thomas Jefferson intentionally changed John Locke's seventeenth-century formulation of the natural rights of human beings from "life, liberty, and property" to "life, liberty, and the pursuit of happiness" among the "unalienable rights" with which citizens were "endowed by their Creator."

In the Murray case, however, "protect the rights of landowners" was a far-from-subtle attempt to scare readers into believing that "someone was coming" for their property.

In fact, both the mineral-rights holder and the surface-rights holder *are* landowners. That being the case, which landowners' rights were in peril? Further, whether fossils belong to the mineral-rights holder, to the surface-rights holder, or to some combination of the two, they belong to parties whose permission must be sought to excavate fossils and who are entitled to share in the proceeds from any sale of those fossils.

What the AAPS call appeared to mean was "protect the rights of surface rather than mineral landowners," a position that may have depended less on the argument that fossils did not belong within mineral estates for legal or philosophical reasons and more on the worry that past deals for the sale of fossils needed protection. It was, that is, predicated on another hypothetical: that, if the "bad" 9th Circuit ruling stood (i.e., fossils were part of the mineral estate), fossils collected, sold, or traded on the basis of agreements with surface-rights holders might suddenly belong to someone else—the mineral rights holder—nullifying previous deals and prohibiting the future sale of fossils already collected on the basis of accords with surface-rights holders. As a result, or so the "slippery slope" argument went, fossils would be confiscated or tied up in lengthy legal disputes over ownership. There was, moreover, the implicit warning that this "threat" would spread across the United States.

In its own opinion, however, the 9th Circuit addressed these fears directly:

> [A] museum's ownership of fossils would only be in doubt following this decision if the museum purchased fossils from the owner of the surface rights of the property where the fossils were found [and] the mineral estate was owned by another party that did not consent to the sale of the fossils to the museum.... Even then, if the mineral estate's owner successfully sued the museum for ownership of the fossils, the museum could recover the value of the sale from the owner of the surface estate. (*Murray and Murray v. BEJ Minerals, LLC, and RTWF, LLC*, 2018, FN 11)

What that means is that the risk that "all fossils" were in danger, or that museums might be forced to give up important fossils, was wildly overblown.

There was, moreover, no indication that the 9th Circuit intended its ruling to apply to all past sales of dinosaur specimens[22] (the fact that the fossils in the case were discovered in 2006 was relevant to the *Murray* case, but had no bearing on other, previous sales), and the *high value* of the fossils played a role in the thinking of every court that touched the *Murray* case. There's no reason to think that the ruling would have been extended to more ordinary or less profitable fossils. Even if it had, however, the point remains: It is immaterial—at least for future exploration and sales—whether the fossils belong to the surface-rights holder or the mineral-rights holder. If they were found on private property, one rights-holder or another would be entitled to profit from them unless she or he had contracted otherwise.

The overriding concern, in fact, was that *private* fossil owners and dealers, who had obtained their fossils through arrangements with surface-rights owners, might not own them (or their profits) anymore. In other words, the issue wasn't the protection of fossil resources or the patrimony of museums but the safeguarding of commerce.

Recall that the issue became a legal contest only because the Murrays attempted to sell the "Dueling Dinosaurs" at auction, and parties who feared being left out of the profits objected. Because the Murrays chose to market the fossils via the auction process, the specimen was just as likely to end up in a museum, in the living room of a movie star, or in the penthouse showroom of an anonymous private collector in Abu Dhabi or Tokyo. Science and the public trust, clearly, were not foremost on the Murray's minds.[23]

When overblown claims about the virtue of fossil sellers are made, then, the reality that high-value fossils often end up somewhere other than a public museum—because they can command much higher prices in the private sphere—needs to be part of the conversation. That doesn't mean all fossil dealers are venal and "only

---

[22] Some fossil dealers disagree with this assessment. What can be said without controversy, however, is that the 9th Circuit opinion was completely silent on the question of whether the ruling was intended to be retroactive. That doesn't mean lawsuits may not have arisen in an attempt to apply the ruling retroactively, but that is a different issue, and any statement about how such hypothetical lawsuits might have been resolved is guesswork. At the same time, the high cost of litigation strongly suggests that such conflicts would have arisen only in cases of extremely valuable fossils, which returns us to the question of the inflated economic valuation of fossils on the very "free market" that some fossil dealers so vigorously defend.
[23] In the end, Dueling Dinosaurs was not auctioned off because bids failed to meet the minimum asking price of six million dollars.

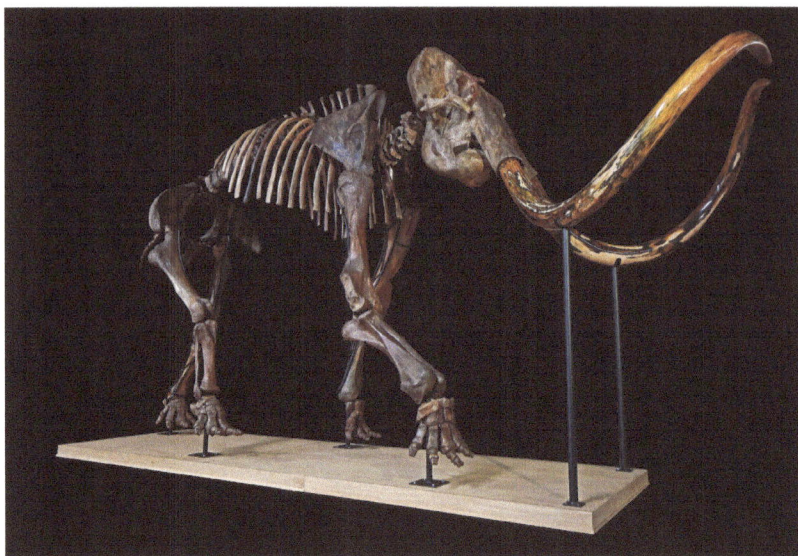

Top: The Wyoming dinosaur sold by Aguttes for the equivalent of nearly $2.6 million in 2018 to an anonymous "French art collector." Below: The "world's largest privately-owned woolly mammoth skeleton," which Aguttes sold for the equivalent of roughly $650,000 in 2017. Aguttes' website proclaims: "In December 2016, Aguttes revolutionized the world market for the sale of large fossils by obtaining a record result of more than one million euros auctioning an *Allosaurus* skeleton.... The adventure continues in June 2018 [with] the auction of a specimen of carnivorous dinosaur ... for €2,019,680 from the first floor of the Eiffel Tower, thus establishing a [new benchmark in] this market.... Our expert Eric Mickeler, member of the European Chamber of Art Consultants, continues to popularize paleontology with the general public.... Along with this extraordinary speciality, our exhibitions and auctions are skilfully orchestrated to highlight the unique and historical nature of the species discovered. Publicised exhibition sites, the opportunity for scientists to study these specimens, the possibility for our customers to name species in accordance with scientific rules, these are the services offered by the department to assist you in the purchase or sale of dinosaurs by auction" ("Dinosaurs & Natural History," 2019). According to Mickeler, "herbivores do not quite excite businessmen who buy dinosaurs the same way as carnivores do. They want to buy carnivores like themselves" ("Rare Dinosaur Skeleton," 2018).

in it for the money," but it does mean that money is a motivation—in some cases, the only motivation. That fact has to be acknowledged not as a slam against the sellers of fossils but as an element of the industry in which such sales take place—and Phipps, the Murrays, and the Seversons were examples of it.[24]

The competition for important and unusual specimens that exists between fossil merchants and private collectors, on the one hand, and, on the other, museums and institutions, is anything but hidden. On the contrary, it takes place in the open:

"The Dinosaur Trade: How Celebrity Collectors and Glitzy Auctions Could Be Damaging Science": "'Fossils are not like ordinary art objects,' says David Polly, president of the Society for Vertebrate Paleontology. 'A skeleton like this [a Wyoming dinosaur auctioned in Paris in 2018 to an anonymous bidder for nearly $2.6 million] is potentially a unique and irreplaceable piece of evidence of earth's past, and in that sense it's important to all of us'" (Reynolds, 2018);

"Dinosaur Fossil Collectors 'Price Museums Out of the Market'": "[P]alaeontologists ... fear not only the loss of scholarship but also the diminution of appreciation for the work that goes into discovery, excavation and

---

[24] In the 2020 Discovery Channel series, *Dino Hunters*, Clayton Phipps lamented that he had yet to sell the Dueling Dinosaurs. Though some paleontologists say the specimen has scientific value, others reject it as useless because of the lack of ancillary data that were not recorded during excavation. *Science* reported that Phipps' failed attempt to auction the Dueling Dinosaurs in 2013 has made "[m]any paleontologists fear that [he] will sell to a private collector who may not allow detailed scientific study. If that happens, 'then someone might as well walk up to it with a sledgehammer and turn it to dust,' says paleontologist Thomas Carr of Carthage College in Kenosha, Wisconsin" (Pringle, 2013).

reconstruction of a dinosaur skeleton as they become fashionable objects of home décor" (Helmore, 2019);

"Dinosaur Skeletons Aren't Décor—They Shouldn't Be Sold to the Highest Bidder": "[W]e've been too lax about letting scientifically important fossils be ushered away into uselessness by those looking for a conversation piece" (Switek, 2015);

"Bones of Contention": "Ranchers who had once allowed scientists to explore their land for free began leasing it to the highest bidder. Paleontologists lost out to amateurs with more money, and they lost specimens to vandals and thieves, some of whom went after fossils with sledgehammers. Federal agents have tracked stolen American dinosaurs as far away as Japan. The paleontologist Kirk Johnson, the director of the Smithsonian's National Museum of Natural History, says, 'The day Sue [the *T. rex*] got auctioned is the day fossils became money'" (Williams, 2013);

"Forget the Old Masters, It's All About the Old Monsters in the Booming Market for Dinosaur Fossils": "[A]s the market roars and collectors rush to snap up the best gems, it may become more difficult to keep these prized pieces of natural history on public view—and within reach of scientists.... But Eric Mickeler, who led the [June 2018 Aguttes sale], says dinosaur fossils are akin to other gems. 'It is a mining product resulting from a fortuitous meeting, similar to the discovery of an exceptional diamond'" (Brown, 2019).

"The Dinosaur Fossil Wars": "The paleo-passion ... extends far beyond celebrities. A money manager in New Jersey, whose office displays several striking fossil specimens, including a three-foot-long Cretaceous *Psittacosaurus,* commented that 'the group who used to be serious fossil collectors—that's really grown. Since the book and movie *Jurassic Park*, interest in fossil collecting has gone into overdrive, affecting demand and elevating prices'" (Webster, 2009);

"Carnivorous-Dinosaur Auction Reflects Rise in Private Fossil Sales": "Paul Barrett, a paleontologist at the Natural History Museum in London, says that because museums are buying fewer fossils, owing to tighter budgets and the fact that many established museums already have large collections, commercial fossil collectors may now be targeting private buyers. 'There seem to be a larger number of wealthy companies and individuals interested in acquiring dinosaurs,' he says'" (Pickrell, 2018).

"Rustlers Finding Gold in Dinosaur Bones": "Sadly," said Michael Woodborne, a University of California paleontologist, ... "these resources, which mean so much to science if properly investigated, are being ripped out of the ground by people who see them not as the gateway to knowledge but as mere trophies." (Coates, 1991);

"Paleontology and Private Fossil Collecting Can Be at Odds in the Hills of Wyoming": "[I]n many cases, Green River's best and rarest specimens—prehistoric horses, crocodiles, birds, and more—are reserved for those with the deepest pockets, destined for kitchen islands and backsplashes.... [I]n the popular world of commercial fossils, cash beats scientific knowledge—and there's lots of cash to go around. As more illegally collected fossils hit online shops and auction blocks, museums are having to buy up significant (and expensive) specimens" (Switek, 2018); and

"Bones of Contention: The Regulation of Paleontological Resources on the Federal Public Lands": "The high-priced market for fossils is making it more difficult for paleontologists to protect legally permitted sites currently under excavation, as well as forcing them to compete with commercial dealers who can outbid them for the rights to work sites on private lands. The potential profit from the sale of fossils can also lead to the inadvertent destruction of delicate fossil remains by inexperienced collectors who see only the dollar signs of profit in the remnants of the past" (Lazerwitz, 1994).

GOING, GOING, GONE

"Selling fossils" may bring to mind roadside rock shops, quiet deals with museum curators, or, in a more updated version, eBay storefronts with crisp color photos of astounding specimens. But national and international auction houses are one of the most important venues for the sale of unusual and high-ticket fossils.

Many well-known auctioneers (Heritage Auctions, Christie's, Bonhams, Aguttes in Paris, Able Auctions in Canada, and others) maintain "natural history" departments dedicated solely to the sale of fossils, minerals, meteorites, and other "scientific" specimens. Literally thousands of fossils are up for auction at any given time, ranging from "meg" teeth and trilobites to complete *Triceratops* skeletons (top price: $657,250.00 in 2011 at Heritage) and ichthyosaurs (top price: $239,000.00 in 2008, also at Heritage; see also Sutton, 2018; and Brown, 2019). Though part of the sales pitch at auction houses is typically that especially pricey specimens are "museum-quality," the fossils almost always go to private buyers.[25]

---

[25] Heritage Auctions was involved in the sale of Eric Prokopi's smuggled Mongolian *Tarbosaurus* in 2012 (see below), and maintains a vast catalogue of fossils and minerals, including Moroccan trilobites, dinosaur bones, and *Spinosaurus* teeth; Burmese amber; oreodont skulls; Oligocene tortoises; fossil whale and horse material; narwhal tusks; fossil (described as "antique" in the catalog) and extant

In 2019, for example, the nearly complete specimen of a theropod dinosaur was sold at auction to a private buyer in Paris for $2.6 million. The skeleton is similar to that of an *Allosaurus*, but "differences in features including its teeth, skull and pelvis ... are significant enough for it to be considered a new species (Pickrell, 2018). David Polly, then President of the Society for Vertebrate paleontology, wrote the auction house asking it to cancel the

tionally, that Aguttes removes all doubts about provenance "by systematically asking for the legal file of the dinosaur [which must include] the title deeds of the land, the excavation authorizations, the paid mineral right, the quarry map, the bones map, the export papers showing the customs taxes paid." Hans Sues, however, curator of vertebrate paleontology at the Smithsonian Institution, noted, "A lot of the fossils that come on the market are from countries

sale, but it did not. Meanwhile, Aguttes spokesperson Eric Mickeler said he hoped the fossil would be placed in a museum, but, however it turned out, "the sale [was] honest, public, legal, and documented" (Pickrell, 2018).

Mickeler went on to tell Artsy.net, a leading online clearinghouse for art auctions and sales nationally and interna-

like China, Argentina, and Brazil that have very desirable types of fossils, including dinosaurs, and even though these countries have regulations on paper, somehow stuff always gets out. You go to any major mineral show and there are dinosaur skeletons, or woolly mammoths, or woolly rhinoceroses—beautiful skeletons that get out of these countries. So obviously, the law and the enforcement of the law are quite different" (Sutton, 2018).

That appears to be true in the case of Mongolian fossils as well, whose export and sale is illegal. Wynne Parry, writing for LiveScience in 2012, revealed that his research had shown that

---

elephant ivory; Messel pit birds; hadrosaur and cave bear skeletons; dinosaur tracks; *Triceratops* and *Allosaurus* skulls; ichthyosaurs, and scores of Green River fish and rays, among hundreds of other items. In 2020, Craig Kissick, Director of Nature and Science for Heritage Auctions, was elected by the AAPS as its president.

[t]he auction house I. M. Chait offered the skull of a *Tarbosaurus* ... for sale on March 24, 2011, and a *Tarbosaurus* leg this past May 6. Skinner Auctioneers & Appraisers in Los Angeles offered two dinosaur eggs from Mongolia's Gobi Desert on June 2. Bonhams offered a *Tarbosaurus* skull on May 17, 2011, and, later in the year, on Dec. 11, offered a frilled *Protoceratops* skeleton found in the Djadokhta Formation, which is located primarily in Mongolia. (The remainder of the formation is in China, which also forbids the export of fossils.) (Parry, 2012)

Morocco has also realized that its paleontological heritage is being siphoned away and has begun beefing up enforcement of its own laws. In 2017, Morocco succeeded in blocking the Hôtel Drouot auction house in Paris from selling a plesiosaur skeleton that had been smuggled out of the country in 2011. The skeleton, seventy-five-percent complete, had been priced starting at the equivalent of about $400,000.00 (Cabrera, 2017). In 2018, Morocco investigated the sale of the tail of an *Atlasaurus*, allegedly from the Atlas Mountains, for $96,000 by the Morton Subastas auction house in Mexico City, Mexico ("Morocco Probes," 2018), and Ahmed Benlakhdim, head of the geology department of the Ministry of Mines' geology department, told *Libération* that same year that a bill to "regulate the extraction, marketing and exportation [of] rare specimens and to establish a national collection" was "in the process of being drafted" (Ollivier, 2018).

To return to the *Murray* case: Repeating, often derisively, that some ill-defined "they" were trying to prove that "dinosaurs are minerals" was no more than diversion and click-bait. It took the focus off the far more important questions, such as whether important scientific specimens found in the U.S. should stay in the U.S., whether they should be bought, sold, or owned by anyone outside of the public trust—and, if they can be, whether any limitations should be placed on that commerce in order to balance competing interests. Whatever one's opinion about these issues, it seems obvious that the opportunity to advance a long-delayed conversation was squandered in the hype over Montana's "Dueling Dinosaurs."

OWNERSHIP & "PRIVATE" PROPERTY

In the midst of the *Murray* case, Mike Triebold, a past president of the AAPS and currently president of Triebold Paleontology Inc. in Woodland Park, Colorado, was interviewed, and one line from his comments circulated widely: "If they don't like the free enterprise system, don't participate in it, but don't demonize those that do" (see, e.g., Pringle, 2014). It's an interesting position—first because it insists on the familiar, categorical, zero-sum battle lines that have shown themselves to be utterly useless up to now: Anyone who raises questions or doubts about "free enterprise" in general or about "free fossil enterprise" in specific is "demonizing" it. And, second, because the corollary of Triebold's statement merits equal consideration: If you are in thrall to the free-enterprise system, don't demonize those who find some of its consequences alarming.

In considering this appeal to the expedients of free enterprise, it's worth keeping in mind some of the other innovations of the free market: the ninety-hour work week; minimum wages far below the cost of living; dangerous and even deadly workplaces; an alarmingly disproportionate concentration of wealth; the offshoring and outsourcing of American jobs; environmental disasters; a severe reduction in food- and product-safety monitoring; crushing student-loan debt; a population of twenty-seven million Americans without health insurance; massive leaps in prices for consumer goods, food, rent, and utilities; a significant decline in the enforcement of antitrust laws and the rise of giant corporate monopolies;[26] destruction of workers' rights; the Great Depression of the 1930s; the deregulation and subsequent failure of the savings and loan industry in the 1980s; and the deregulation of banks and resulting real-estate-market crash in 2008. "Free enterprise," moreover, is all too often shorthand for "free of regulation," the neoliberal, laissez-faire, "invisible hand" approach to business, which has, in fact, proven to be incapable of protecting the good of the many in virtually every context in which it has been applied.

A related philosophy, expressed by Peter Larson in a series of emails to *Fossil News* about the *Murray* case in late 2018, goes hand in hand with a belief that "free enterprise" holds a priority over all else. In the context of a discussion about whether specimens of great scientific value should be privately owned or rather held in institutions for the public trust, Larson commented: "The last I heard, private land is just that. Check out the US Constitution if you think you can nationalize fossils without violating the principles of private ownership."

First of all, carving out a distinction for especially scientifically valuable fossils isn't synonymous with and doesn't even require "nationalizing" fossils. More importantly, private ownership is a social relationship among citizens, not merely a transaction between a person and a thing. Inherent in it is the power to exclude, by force and even violence ("Trespassers Will Be Shot"), and that's worth some honest reflection.

Second, the Constitution doesn't confer (or deny) a right to hold property. Private property wasn't even mentioned in the original document, and appeared only when the Bill of Rights was adopted in 1791, three years after the

---

[26] See, e.g., Stucke & Ezrachi (2017) for a chronology of antitrust laws in the U.S.: "[If the country] were to continue with a "light-if-any-touch" antitrust review of mergers and a blind eye to abuse, concentration will likely increase, our well-being will decrease further, and power and profits will continue to fall into fewer hands.""

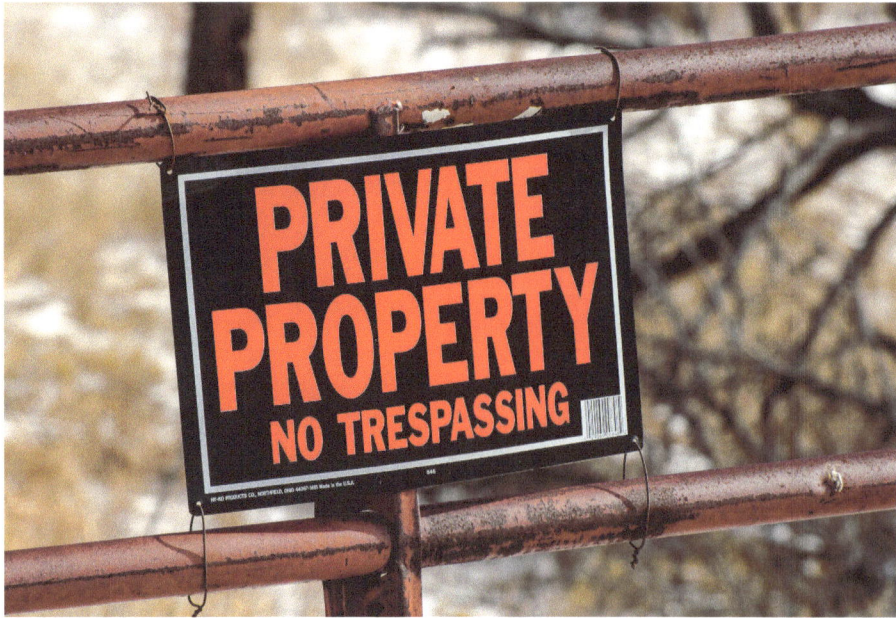

of dispossession of resources from indigenous lands.[28]

But the topic is fossils, and the point is not to litigate the merits of the free vs. regulated market or to advocate the abolishment of private property. Rather, it is to ask why the "free enterprise" system in the fossil industry deserves so much reverence with respect to all other questions of equity and why private land ownership merits such a heated defense, especially given its historical context in this country and particularly when ownership of real property is not now equitably distributed and is unavailable to a significant proportion of Americans.[29] Why, in other words, are these so often the first lines of defense for the commercial fossil industry?

Or, to put it another way, is there really no choice but to frame issues of fossil commerce within these terms alone? They need not be banished from the conversation, surely, but neither must they be the entirety of the dialogue or the wellspring from which all policy and debate must flow.

Constitution was ratified. What the Bill of Rights says is that the government cannot deprive anyone of property without "due process of law" or take private property "for public use without just compensation." This means that government can sometimes exercise the power to operate in the public interest (the "Takings Clause" of the Fifth Amendment), subject to limitations recognized by the courts. The Fourth Amendment's ban on "unreasonable searches and seizures" protects private property from "invasion."

Private property (in this context, meaning chiefly real estate) is nonetheless commonly "taken" to benefit the public interest (to build parks or roads, for instance). This is not to argue that government bodies are uniformly ethical or principled, that the compensation they offer is always "just," or that "due process" is invariably observed, but it is to say that the Constitution provides for the possibility far, far short of "nationalization." Moreover, if a landowner is excavating fossils (or allowing them to be excavated) in order to sell them, what difference does it make if the specimens are bought by a private individual or if the government "justly compensates" the owner in order to keep the fossils in the public domain?

There is something willfully blind to history, meanwhile, in the insistence that the concept of "private property" is untouchable in the context of land that was stolen from the people who originally lived there[27] and of the long history

DINO HUNTERS

Other aspects of the image of commercial fossil collection that the AAPS promotes include the commercial fossil hunter who is "as good as any academic" when it comes to field methodology or scientific knowledge and who, though he may be justifiably trying to "make a living," isn't "in it only for the money." If those are important precepts for the organization, the opportunity to educate the public should have been irresistible when the Discovery Channel premiered its new series,

---

[27] Historian Jeffrey Ostler notes that the twenty-seventh and final grievance against Britain in Jefferson's Declaration of Independence was that the Crown had "excited domestic insurrections amongst us" (meaning slave revolts) and brought "the merciless Indian savages" down on the colonists (at the time, indigenous people, supported by Britain, were violently resisting the invasion of their lands west of the Appalachian Mountains) (Ostler, 2020).

[28] The intentionally provocative comment is often made, in this context, that indigenous Americans "had no concept of private property," one among Ayn Rand's more notorious claims. Apart from the fact that generalizations about the cultural and legal structures of hundreds of bands, tribes, and communities are rarely useful, a somewhat more accurate statement might be that, unlike Europeans and Euro-Americans, Native Americans didn't traditionally view land as an economic asset, and land did not typically confer power or status on individuals.

[29] According to the U. S. Census Bureau, 35% of Americans do not own homes in 2020—26% of whites, 56% of blacks, and 51% of Hispanics. These statistics have changed little over the last two decades. If the issue is land, the top 10% of households owned 82% of the country's non-home real estate in 2017, according to the *Washington Post* (Ingraham, 2017).

Dino Hunters, in June 2020. Indeed, for its portrayal of commercial fossil collectors and of the fossil market, *Dino Hunters* was everything the AAPS says it disdains.

Among the more general fossil-interested public, meanwhile, *Dino Hunters* became the subject of calls for boycotts almost from the moment the series appeared. The hashtag #CancelDinoHunters was born on Twitter and Instagram, and social media paleontology commentators couldn't find enough ways to malign the program. As one poster wrote on Instagram, "There are NO paleontologists in this show and no ethical science conducted—just a bunch of greedy & corrupted clowns looking to profit off fossils & promote fossil commercialization." Another wrote, "*Dino Hunters* goes against everything that scientists [believe]," and still another urged others to respond to *Dino Hunters* by "informing [yourself] on why this [series] in no way represents paleontology or what it stands for" and by "promot[ing] good fossil hunting ethics and shar[ing] any important fossil finds with local museums, paleo societies or professionals."

One might object to *Dino Hunters* on the grounds that it offers the same fake-drama, low-content pseudoscience that the Discovery Channel has been churning out for years, but a particularly galling aspect was the show's blatant and ultimately tedious message—that "Dinos Make Bank," as one of the series' promo spots put it. The entire context of *Dino Hunters*, in fact, which spotlights various groups of "hunters"—Clayton Phipps of Dueling Dinosaurs fame is the star of the "Montana" segment; Mike and Jake Harris and John and Aaron Bolan appear for Wyoming; and Jared Hudson is shown at work in South Dakota— is that commercial fossil hunting is driven almost exclusively by economics. Each episode begins with this dramatic introduction:

Discovery's "Dino Hunters": Never at a loss for cowboy hats. Don't try collecting fossils without one.

Across the American west, ranchers are fighting to preserve their way of life, struggling to make ends meet by raising cattle. A few modern-day frontiersmen have found a new way to save their livelihoods—hunting for dinosaurs.

Every single fossil excavated on the series is immediately assigned a dollar value, often before any scientific information has been provided, and the show's "frontiersmen" and "cowboys" talk endlessly about money—how much excavations cost, how close they are to being unable to "feed their families," about the need to "keep their bankers happy" or the financial disaster that will befall them if anything goes wrong, about what they can sell quickly "to make ends meet" or about "not being able to finish out the season if we don't find something soon."

Concerns about money are common to Americans, and with good reason, but the harping on the market value of fossils ultimately becomes hard to reconcile with the idea that commercial collectors do what they do for the love of fossils and not solely "for the money."

In the first minutes of the first episode, in fact, Clayton Phipps utters a glaring falsehood. Talking about the high asking prices of complete dinosaur fossils, he says, "It's museums that can usually raise the cash for the big skeletons." In fact, it is not; it's private collectors, celebrities, and corporations. (The $8.36 million that the Field Museum paid for Sue the *T. rex* was entirely provided by "McDonald's Corporation, Ronald McDonald House Charities, Walt Disney World Resort, the California State University System, and private individuals; Browne, 1997).

Perhaps that remarkable sale is what led John Bolan to explain, later in the series, his "financial decision" to abandon a fruitless dig for what he had hoped would be a complete hadrosaur. He and his team would be better compensated by selling the site to a museum, he opined, adding that "A lot of museums, they have no real risk. They have all the time in the world, and a third party is paying them for this." Again, if it needs to be said, that is hardly ever true. The Field Museum's unprecedented (before and since) fundraising effort was for an exceptional celebrity dinosaur, and both Disney and McDonald's received more than their money's worth in publicity (and sales) in return for displaying and touring with life-sized casts of Sue. That's not even remotely in the same ballpark as paying a team to excavate a "maybe" hadrosaur.

By definition, meanwhile, museum funding is virtually *always* at risk because very few museums in the U.S. are fully publicly funded. According to one study, museums in the U.S. receive government support (from any level of government) at levels that "ranged between 7% and 33% by museum type... [W]hen government administered museums are removed from the analysis, the highest proportion of public support drops to 24%" (Manjarrez et al., 2008: 8). Non-profit natural history and natural science museums receive 13.9% of their funding from government sources (27). The Manjarrez group's data are twelve years old, however, and every indication is that state and federal funding for museums has declined significantly and steadily since they were gathered.

As social media critics of *Dino Hunters* noted, in addition, no accredited paleontologist is involved in the show's excavations nor is one ever interviewed (one individual, whom the narrator identifies as a "paleontology expert," is an expert preparator but his only degree is a bachelor's in biology). When scientific information is provided, it is often wrong: *Nanotyrannus* is no longer considered a valid species,   except by the

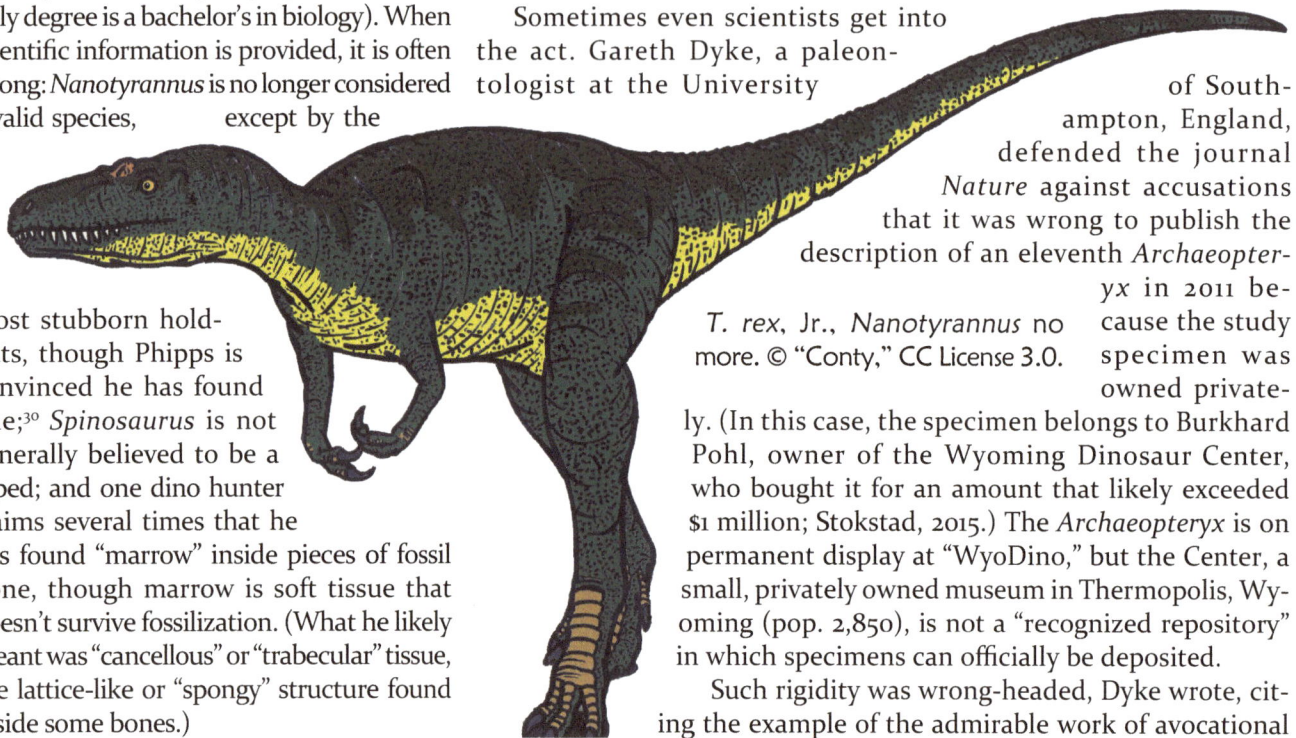

most stubborn holdouts, though Phipps is convinced he has found one;[30] *Spinosaurus* is not generally believed to be a biped; and one dino hunter claims several times that he has found "marrow" inside pieces of fossil bone, though marrow is soft tissue that doesn't survive fossilization. (What he likely meant was "cancellous" or "trabecular" tissue, the lattice-like or "spongy" structure found inside some bones.)

No one ever discusses, let alone collects, information on the stratigraphy, position, taphonomy, associated fossils, or paleogeography of the sites. Nor, finally, are important discoveries reported to any institution or museum,[31] despite the series'

claim, in its online promo materials, that the "groundbreaking finds" by the "dino hunters" "[have] set the scientific world on fire."[32] Perhaps other of their finds have done so, but "the scientific world" appears never even to have seen the fossils that Phipps, the Harrises, the Bolans, Hudson, Murphy, and others excavated on *Dino Hunters*.

PROOF BY EXAMPLE: NOT MUCH PROOF

In public materials in support of commercial collecting, one common AAPS device is the logical fallacy known as "proof by example"—the attempt to demonstrate the validity of an assertion by citing a series of individual cases. The organization's "Fossil Specimens Placed in Museums and Universities by Commercial Paleontology" page is an example, as is the claim that "magnificent displays" of fossils grace "any number of prestigious institutions" thanks to professional collectors (Triebold, 2007).

The point always seems to be to create a perception that is totalizing. If one careful, science-minded commercial collector exists, then they all are. If one fossil dealer has generously donated to a public museum, then they all do. If one snobbish, anti-commerce paleontologist exists, then that's true of the rest of them, too.

Sometimes even scientists get into the act. Gareth Dyke, a paleontologist at the University of Southampton, England, defended the journal *Nature* against accusations that it was wrong to publish the description of an eleventh *Archaeopteryx* in 2011 because the study specimen was owned privately. (In this case, the specimen belongs to Burkhard Pohl, owner of the Wyoming Dinosaur Center, who bought it for an amount that likely exceeded $1 million; Stokstad, 2015.) The *Archaeopteryx* is on permanent display at "WyoDino," but the Center, a small, privately owned museum in Thermopolis, Wyoming (pop. 2,850), is not a "recognized repository" in which specimens can officially be deposited.

Such rigidity was wrong-headed, Dyke wrote, citing the example of the admirable work of avocational collectors on the Isle of Wight, "one of the best places in Europe to collect the remains of dinosaurs," who have discovered new species and whose private collections "can tell us a huge amount about ... early Cretaceous ecosystems." Collaboration was the key, Dyke ar-

T. rex, Jr., *Nanotyrannus* no more. © "Conty," CC License 3.0.

---

[30] See Woodward et al. (2020). Growing up *Tyrannosaurus rex*: Osteohistology Refutes the Pygmy "*Nanotyrannus*" and Supports Ontogenetic Niche Partitioning in Juvenile *Tyrannosaurus*. *Science Advances, 6*(1). DOI: 10.1126/sciadv.aax6250.

[31] Recall that the AAPS's Code of Ethics

states: "5. Report to scientific experts any significant discoveries of scientific or public interest; 6. Strive to place specimens of unique scientific interest into responsible hands for study, research and preservation." Interviewed about the Dueling Dinosaurs in 2013, Mike Triebold told the Colorado-based *High Country News*, "'It's in the best interest of commercial companies to be stringent data collectors so they can sell to quality museums.' He believes data collection by commercial operations can be superior to that done by academic institutions strapped by budgets and bureaucracy" (Hodges, 2013). It *can be* but, as *Dino Hunters* demonstrates, it isn't *necessarily*.

[32] See https://www.discovery.com/shows/dino-hunters.

gued, because "the science of paleontology cannot exist in a professional vacuum" (Dyke, 2015).

That anyone is in favor of "professional vacuums" or opposed to "collaboration" is doubtful, but here is where apples and oranges collide. None of Dyke's anecdotes are relevant to the legitimately held opinion that important specimens should reside in public repositories of a specific kind. Debates can (and do) take place over that professional standard, but it isn't automatically rendered incorrect because virtuous amateur collectors exist.

Journalists who write about fossil commerce fall into an identical trap, perhaps hoping, if they make no challenge to what they are told, that their work will seem more objective. Lewis Simons' "Fossil Wars" for *National Geographic* is one example among hundreds:

> I spent months tracking commercial fossil dealers and investigating their trade, not just in Siberia and Colorado, but also in Morocco, northeastern China, Montana, and the Dakotas. I discovered that some dealers are careful collectors and honest businessmen; others are disreputable and brutish, ripping bones from national parks and other protected lands and selling them for a quick buck. Still others, particularly in developing countries such as China and Morocco, are peasants striving to ease their painful lives.... . During my travels I witnessed some of the damage that unscrupulous or untrained dealers do. In northeastern China I watched pick-swinging farmers hack rock slabs containing the remains of ancient birds and fish with little more concern than they gave to plowing their fields. I saw smuggled and fake fossils sold as legitimate in the United States, which strictly prohibits the excavation and export of fossils from government-owned land without a permit, but has no law banning imports—even when they've been smuggled out of their originating country. I also watched commercial dealers excavate fossils with exquisite care, cleaning away the detritus of eons ... and keeping finely detailed records of their discoveries. (Simons, 2005)

In summary: Each observation cancels every other observation, so the net impact is neutral.[33]

Still, it's not that claims about virtuous collectors (commercial or not) are false or manufactured. They aren't, and there are plenty of good guys. Richard Conniff, author of *House of Lost Worlds: Dinosaurs, Dynasties, and the Story of Life on Earth,* writing in *National Geographic* in October 2019, described the way that discoveries by

amateur and commercial collectors in the U.S. and abroad "almost inevitably oblige collectors and paleontologists to work together," noting "the extent to which private collectors, commercial fossil hunters, and museum paleontologists now quietly cooperate," in part because of necessity. Kirk Johnson, director of the Smithsonian National Museum of Natural History, told Conniff that "[c]ash-strapped museums everywhere have cut research staff and budgets. Commercial collectors are thus 'digging much more than scientists'" (Conniff, 2019).

John Pickrell (author of *Weird Dinosaurs*) wrote in *New Scientist* in 2020 about a sort of paleontological Robin Hood, François Escuillié, the owner of Eldonia, a fossil dealership in France, who has made a practice of buying black-market specimens and donating them to museums, including the holotype of a waterfowl-like dromaeosaurid theropod, *Halszkaraptor escuilliei*, named and described in 2017 by paleontologists Andrea Cau, Khishigjav Tsogtbaatar, Philip J. Currie, and their colleagues, which had been illegally collected and smuggled out of Mongolia some time prior to 2015. The fossil was scheduled to be repatriated to Mongolia in 2020.

Australian rancher Ross Fargher has spent some thirty years protecting the rare Ediacaran fossils that crop out in such profusion near his cattle station in South Australia, and he has created a unique research opportunity for paleontologists, including Mary Droser (University of California at Riverside) and Jim Gehling (South Australian Museum), who have worked on the fossils for nearly two decades. According to Droser, Fargher's commitment—he chases off looters, protects their research sites, runs tours, arranges logistics for visiting scientists, and uses his earth-moving equipment to help expose fresh collecting areas—has allowed them to "invent ... a new way of doing paleoecology" (Finkel, 2019). In March 2019, the South Australian government purchased a large section of Fargher's land, increasing the size of the existing Ediacara Conservation Park by a factor of ten and insuring that his fossils will be protected permanently. In 2018, the U.S.-based Paleontological Society gave Fargher its Strimple Award for amateur paleontology (Finkel, 2019).

In India, meanwhile, where fossil collecting and commerce are all but entirely unregulated, Vishal Verma, a high school physics teacher, has spent twenty years traveling his country to collect and save dinosaur nests, a fossilized gymnosperm forest, and hundreds of other specimens from looting and commercial over-exploitation. Guntupalli V. R. Prasad, a paleontologist at the University of Delhi, applauds Verma's efforts: "He does more field work than any paleontologist," he said (Kumar, 2018: 24).

But anecdotes are not evidence, if evidence is needed, and, standing alone, they cannot serve as the basis for reasoned argument or for public policy. But "proof by example" is a common recourse: if cases can be cited of fossil dealers who care a great deal about science, who

---

[33] Neutrality is when one person says it is raining, another person insists it is not, and the journalist reports both assertions. Objectivity is when one person says it is raining, another person says it is not, and the journalist pokes his head out the window to see whether he gets wet.

are generous in donating to museums, and who regularly partner with professional paleontologists; and if museums contain specimens that wouldn't be there if commercial dealers hadn't found or provided them, then any criticism of fossil commerce must be invalid.

Another weakness of proof by example is that it works both ways. If we're citing examples, in fact, there are just as many, and perhaps more, instances of egregious, unethical, and illegal behavior on the part of private and commercial fossil collectors than there are instances of virtue.

Virtually every national park, national monument, and protected area in the U.S., for example, has a problem with poaching—that is, with visitors taking home whatever they can pick up, including artifacts, stone, wood, rare cacti and other plants, and, of course, fossils. According to a 1998 review of laws protecting fossils on public land, "Due to the expanding commercialization of fossils, it has become increasingly difficult to protect fossils on public lands. To meet the growing demand for fossils, people are stealing them from public lands in the United States, and abroad" (Lundgren, 1998).

There isn't always a way to be sure that fossils are poached in order to be sold—although, in others, the connection between theft and commerce is irrefutable—but it's not unreasonable to think that the proliferation of the fossil trade helps fuel the idea that fossils are *highly valuable*. On this point, experts agree, and fossil dealers themselves can hardly deny that the *value* of fossils (perceived or actual) is what makes their businesses possible. Thefts of fossils from public and private sites happen precisely because some collectors cannot tolerate the idea that a pile of money is just sitting there "when I could have it."

A few specifics:

- The Fossil Cycad National Monument, home to one of the world's greatest concentrations of cycadeoids, a major group of Triassic seed plants known as Bennettitales, was established in South Dakota in 1922. According to Tony Martin of the Paleontological Society, "By 1929, an NPS report on the monument stated that all cycadeoid specimen previously exposed had been removed, and there was 'nothing left... of interest to visitors'" (Martin, 2019).

- "Operation Rock Fish" in Wyoming in 1995-1996 used aerial surveillance of the Fossil Butte National Monument and other public lands to record fossil theft, resulting in twenty-nine felony arrests (Abel, 1996) and the first felony conviction for fossil theft in U.S. history for the theft of fossils from Badlands National Park (Santucci, 2020).

- National Park Service paleontologist Vince Santucci said that, while doing graduate research, he "soon learned that an escalating commercial fossil market was resulting in increased incidents of fossil theft on public lands including national parks" ("Vin-

Dickinsonia, one of the most familiar Ediacaran fossils, from the Flinders Ranges in South Australia, where Ross Fargher's ranch is located. © Flinders Ranges Ediacara Foundation.

cent L. Santucci," 2020). In 1991, the FBI recruited Santucci to investigate fossil theft from protected areas (Santucci was part of the FBI team that raided the Black Hills Institute of Geological Research in South Dakota in 1993, from which Sue the *T. rex* was confiscated; Breznican, 1997).

- In 2017, a trackway of fossilized footprints were excavated and stolen from Death Valley National Park (Martin, 2019).

- Estimates are that some twelve tons of petrified wood are taken without permission from Petrified Forest National Park in Arizona each year (May, 2017).

- The site of extraordinary find of the skull, jaws, and teeth of a 330-million-year-old shark in the Mammoth Cave National Park in November 2019 has been kept secret from the public and cannot be viewed because of "theft and vandalism in the past" (Austin, 2020).

- According to the Nebraska National Forest, more than 20% of the 30,720 acres in the Oglala National Grasslands has been subjected to illegal vertebrate fossil collection; of thirty-nine sites "designated as having special importance because of exceptional preservation of fossils, 11 (28%) showed evidence of unauthorized collecting" (cited in Lazerwitz, 1994).

- In 1991, the Bureau of Land Management (BLM), in cooperation with the Museum of the Rockies, rescued an *Allosaurus* ("Big Al") being excavated from public land by an overseas commercial collector" (Lundgren, 1998).

Other countries have problems with fossil theft, too, and the attempts are often even more daring. In 2015, a man who attempted to sell smuggled Chinese dinosaur fossils at the Tucson Gem & Mineral Show, an annual two-week affair that is the largest fossil market in the United States, was fined $25,000 after being caught by undercover agents (Proctor, 2016). Nearly twenty years earlier, in 1996, the Tucson trade show was the site of one of the largest fossil seizures in U.S. history when federal authorities confiscated four tons of valuable and unusual fossils, including dinosaur eggs, petrified pine cones, and fossilized crabs, all of which had been illegally imported for sale from Argentina (Pagan, 2008).[34]

An Irish fossil dealer, George Corneille, was investigated in 2017 after he advertised a partial *Spinosaurus* jaw from Morocco for sale through his Facebook page. Corneille insisted that *Spinosaurus* fossils are "not rare" and that "dozens [of specimens] can be found in other areas, such as Europe and America" and claimed he had bought the fossil legally at the Tucson show from a Moroccan dealer, though no documents regarding provenance or the sale were apparently ever produced. He was never charged and continues to sell *Spinosaurus* and *Igdamanosaurus* (*Globidens*) teeth and jaws, as well as other exceptional fossils from all over the U.S. and Europe, on his website.

The famous Burgess Shale fossil beds in British Columbia, Canada, are the source of a "fairly active" black market in fossils, according to Yoho National Park Warden Supervisor, Jim Mamalis. "The value can range from $300 or $400 for a fairly common trilobite fossil, but we've seen some of the more rare fossils from that area advertised for sale online for up to $10,000," Mamalis said. In 2016, a Belgian tourist was fined $4,000 for walking off with a trilobite from Yoho that he had hidden in a sock (Nieoczym, 2016).

Between 1991 and 2001, the collection of the Paleontological Institute at the Russian Academy of Sciences in Moscow (Палеонтологический институт or "PIN") was looted of "hundreds of unique fossils, potentially worth millions of dollars. According to accusations, the museum's own directors and scientists set up private companies to sell the fossils and forge the export documents necessary to take those fossils out of the country (Weir, 2001).

Larissa Doguzhayeva, a researcher at the museum and an expert on ammonites, blew the whistle on the practice, explaining that she first became suspicious in 1996 when Arkady Zakharov, at the time a PIN scientist, brought a German fossil dealer to see her with an offer "to buy a big collection of ammonites I had just gathered in field work. I told them it was state property, and asked them to leave my office." Later, she realized the collection had been stolen from the museum, though PIN's director refused to report the incident to the police. The same German dealer was arrested three years later while trying to "transport a truckload of partially undocumented fossils into Finland." By 2001, Doguzhayeva acknowledged that the pillaging had probably ended because the fossil dealers had cut out the middleman: "Russia's rich fossil grounds are being ruined by ruthless predators, working in league with rogue scientists, who smash the sedimentary strata with machinery and cart fossils away by the truckload. 'The last time I went to search for ammonites, near Shilovka on the Volga River, I was stunned by the destruction,'" Doguzhayeva told *The Christian Science Monitor* (Weir, 2001).

In January 2001, Zakharov, who formed the company Russian Fossils when he left PIN, was stopped as he attempted to ship a collection of 12,000 ammonites from the Volga region out of the country; Zakharov had officially declared that the fossils were "scientifically worthless" in order to circumvent export regulations. Zakharov insists the issue of theft or of unchecked over-collecting was invented "by political forces who want to discredit capitalism." As for Doguzhayeva, he says, "she is against business" (Weir, 2001).

One of the most egregious—and most fascinating—cases of "fossil madness" was the subject of Paige Williams' 2013 article in *The New Yorker* and her later book, *The Dinosaur Artist: Obsession, Science, and the Global Quest for Fossils* (2018a). The book's anti-hero was Eric Prokopi, who, as prosecutors in New York described him, operated a "one-man black market in prehistoric fossils." Prokopi had been involved for years in a smuggling operation that poached millions of dollars of fossils, including specimens of the Cretaceous tyrannosaurid, *Tarbosaurus bataar*, from Mongolia.

Prokopi, originally from Florida, had begun selling fossils in high school, and he opened a business, Florida Fossils, shortly after leaving the University of Florida with an engineering degree. At that point, Prokopi began calling himself a "commercial paleontologist." At first, he sold mostly smaller fossils—shark and *Tyrannosaurus* teeth, mammal jaws, partial skeletons, and the like, in addition to doing prep work—but, as Williams tells the story, he realized his

best shot at big money was dinosaurs. The top sites in the American West were largely tied up in federal

---

[34] As early as 1994, the *New York Times* reported that the Tucson Gem and Mineral Show was the site of a federal investigation aimed at "crack[ing] down on fossil trading" and described the fair as "a trade show steeped in dispute, bitterness and litigation" (Browne, 1994, C1). Dr. William S. Clemens, a paleontologist at the University of California at Berkeley, told the *Times*: "I've been visiting the exhibits at the Tucson show this year to get a feeling for the trade, and some of the things I saw made me sick. I saw some exhibits marked with numbers similar to those used by museums, and I couldn't help wondering whether these specimens had been looted from museums. I saw a rare amphibian fossil from Russia on sale, accompanied by a certificate from Russia's Paleontological institute allowing export of this treasure. The Russians must certainly be hard up to let things like that go" (c1, c9).

The holotype of Tarbosaurus bataar, described by Evgeny Maleev, a Soviet paleontologist, in 1955. It is held at the Museum of Paleontology, Moscow (PIN).

As a journalist, Williams is enviably neutral, and *The Dinosaur Artist* is a fascinating international detective story, but it's also an important look inside the world of commercial fossil collecting and selling and the laws that regulate them—or fail to regulate them. At the same time, names that are familiar in the fossil trade today crop up in the story, including individuals who directly or indirectly did business with Prokopi or who had previously sold Mongolian fossils.

The objects of contention are different, but Susan Orlean's book, *The Orchid Thief: A True Story of Beauty and Obsession,* covers some of the same emotional and psychological territory. *The Orchid Thief* is an investigation of the events that led to the 1994 arrest of John Laroche for stealing endangered ghost orchids from the Fakahatchee Strand State Preserve, in the Big Cypress Swamp area of the Florida Everglades, in order to sell them to collectors and horticulture dealers. (Interestingly, Laroche had also been a collector and dealer of Ice Age fossils in his youth.) Laroche initially hoped to be acquitted because of a loophole in the laws, but he was ultimately convicted and received a sentence of six months' probation. He had reasoned, he told Orlean, that "*[s]omeone* is going to figure out how to benefit from the law the way it is now and I just figured it might as well be me…. The state needs to *protect* itself…. I'm really on the side of the plants" (Orlean, 2018: 30).

boundaries and, on the private side, in existing contracts that usually required sizable cuts to ranchers. Only one place on Earth holds big, beautiful *T. rex*-like dinosaurs in relatively soft sand, in a vast, remote landscape that all but insures privacy." That place was Mongolia. (Williams, 2013)

Prokopi's undoing began in 2012 when he sold a *Tarbosaurus* skeleton through Heritage Auctions in New York—to an anonymous phone bidder—for just over a million dollars. What Prokopi didn't know was that a Mongolian paleontologist living in New York had seen the specimen in the auction house's catalog and set into motion a chain of events that would eventually involve Interpol, the federal authorities, and the president of Mongolia. The *Tarbosaurus* skeleton was seized from the auction and eventually repatriated to Mongolia in May 2013, just about a year before Prokopi was convicted on charges that carried a maximum prison sentence of seventeen years. He was given three months ("US Fossils Dealer," 2014). Among Prokopi's defenses was that *Tarbosaurus* fossils were common, anticipating by four years the excuse George Corneille would use during the investigation into his attempt to take a *Spinosaurus* jaw out of Morocco.[35]

Craig Childs' *Finders Keepers: A Tale of Archaeological Plunder and Obsession* also explores these themes in the context of the looting and theft of human artifacts in the American west, but for every instance of the word "artifact" or "antiquity" in the book, it's possible to substitute "fossils"; for every potsherd or 10,000-year-old woven basket, substitute a *Tyrannosaurus* tooth, an intact false sabretooth skull or tortoise shell, a mammoth tusk.

"Diggers [for antiquities] who sell have one thing in common," Childs writes,

---

[35] *Tarbosaurus* skeletons are not, in fact, common, but the sale of Mongolian fossils prior to the Prokopi case evidently was: "Other [*Tarbosaurus*] fossils … as well as other specimens linked to [Mongolia], are not difficult to find in auction

---

house catalogs or on eBay…. A search for 'dinosaur fossils Mongolia' on eBay Wednesday (June 27 [2012]) yielded about 20 results, most of them claws" (Parry, 2012). Update: A July 2020 eBay search for "dinosaur fossils Mongolia" resulted in twelve hits. These were largely eggshell fragments but including two lots of intact fossil dinosaur eggs. A search for "fossils Mongolia" returned more than seventy items, most of them invertebrates and most of them offered for sale "from China." The provenance of an additional seven results, including five lots of dinosaur eggs, a *Psittacosaurus* skeleton, and a *Psittacosaurus* skull, was given as "East Asia."

They dig because there is demand. Antiquities are one of the top illegal trades in the world.... On the world stage, artifacts from the American Southwest make up a dependable but relatively backwoods commerce. Meanwhile, diggers in Central America are trenching into monumental city-states rather than villages, and tombs instead of graves.... The money is much bigger, and so is the scale of looting. (95)

In the U.S., threatened plants and indigenous artifacts are, legally at least, not so different from fossils. Endangered plants are covered by the Convention on International Trade in Endangered Species of Wild Fauna and Flora (CITES), and can't generally be collected, removed, sold, or exported from public land, but only a few states extend those protections to private land. Under the Archaeological Resources Protection Act and

the National Historic Preservation Act, archaeological sites and artifacts are fully protected on federal land as natural and cultural resources, but little guidance exists regarding private property. Some archaeologists think these laws don't go far enough and ought to extend to "*all* archaeological artifacts; no matter ... whose property they're on" (Adams, 2018). The acts mentioned above exclude human remains and burial offerings, which are dealt with under a different law, the Native American Graves Protection and Repatriation Act, but the NAGPRA applies to private property only when a gravesite is excavated or disturbed. If it is not, no one can force a property owner to allow it to be excavated (Adams, 2018).

Fossils are treated somewhat similarly but through a different set of laws: the Paleontological Resources Preservation Act. Casual, non-commercial collecting is generally allowed on public land (with exceptions that are arguably vague and bureaucratic), and other collecting by qualified individuals "for the purpose of furthering paleontological knowledge or for public education" (meaning, generally, professional paleontologists) is allowed by permit only.[36] Almost all of the provisions of the PRPA have been a sore point with some scientists and some avocational and commercial collectors for more than a decade, and various proposals have periodically been offered for revising it. So far, they all have sunk in a hurricane of squabbling. The regulations of the National Park Service, the U.S. Forest Service, the Bureau of Land Management, and, potentially, other federal land-management agencies, interpret the PRPA, though not always in the direction of greater clarity.

In any case, the PRPA cannot be enforced on private property if collecting is carried out by the owner or anyone to whom she or he has granted permission. Nowhere in all of this does one find the concepts that fossils could be endangered or threatened; that they are natural, cultural, or scientific resources; or that collecting might have an impact upon the environment.

Williams, Orlean, and Childs, meanwhile, all explore the conviction, so deeply embedded in America's self-mythology, that the individual who discovers something "in nature," and then thinks of a way to exploit it, is doing no

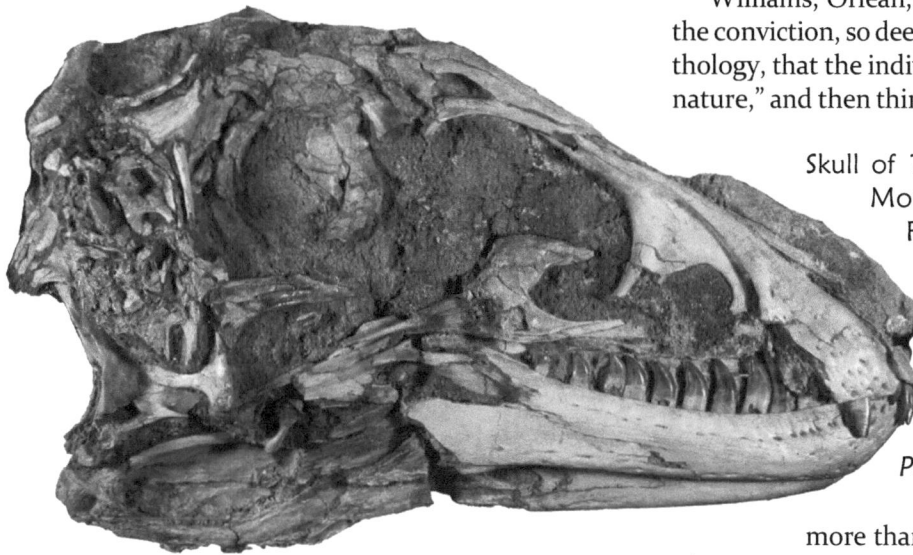

Skull of *Tarbosaurus bataar* housed at the Mongolian Paleontological Center. From Tsuihiji, T. et al. (2011). Cranial Osteology of a Juvenile Specimen of *Tarbosaurus bataar* (Theropoda, Tyrannosauridae) from the Nemegt Formation (Upper Cretaceous) of Bugin Tsav, Mongolia. *J. of Vertebrate Paleontology 31*(3), 1-21.

more than fulfilling his natural (and national) destiny. Sometimes, it's just about bragging rights: I found something rare; I found something worth money; I have something other people covet. More often, it's a reflection of the imperialist mentality that runs like a coal seam through debates about ownership, control, and exploitation of fossils: If I found it, or if I can manage to get control of it, or if I can assert the most compelling claim to it, then it is mine. In other words, finders keepers.

Adopting that playground taunt as the title of his book, Craig Childs traced the ethical, legal, and moral landscape in which debates take place over who should have the right to find and collect artifacts (archaeologists vs. amateur diggers), where antiques should be curated (private collections, tribal headquarters, public museums), and who has the right to have access to them (the public, universities, accredited researchers).

Every detail of the venomous arguments he describes over the "right" to have access to artifacts is identical in

---

[36] A .pdf version of the Paleontological Resources Preservation Act is available at https://irma.nps.gov/DataStore/DownloadFile/640026.

all but the particulars to the collisions among accredited paleontologists, fossil dealers, and avocational collectors: every anecdotal claim that amateur collectors and professional dealers are science's most faithful friends; every "no one can tell me what I can do with what I find," regulation-flouting pot digger; every "respected businessman," "good ol' guy," or "local neighbor boy" who turns out to be involved in an international smuggling operation; every action that the perpetrator justifies as "not hurting anyone"; every instance of disdain that private collectors and antiquities dealers heap upon "elitist" academics and "the government" is duplicated in the world of fossils, as

of fossil and mineral specimens (Sicree, 2009). The trade reportedly brought $40 million into Morocco each year—in 2000 (Osborne, 2000).

These unlicensed and unregulated Moroccan excavators may originally have entered the fossil trade through selling the odd specimen to tourists, but, in the last two decades, fossil commerce has become a major industry with its own supply chain: at the bottom are local workers who sell to wholesalers who, in turn, sell to exporters, some of whom can earn the equivalent of $100,000 a year. (Ben Yahia, 2019). According to mineralogist Andrew Sicree, in the rural areas from which fossils often come,

Above and facing page: Fossils on sale at the Tucson Gem & Mineral Show. CC License 2.0.

is every bit of condescension from science professionals who disdain private collectors because they allegedly spoil archaeological sites, are unscientific and "unprofessional," or because they're "only on it for the money."

Williams, then, does for paleontology what Childs did for archaeological artifacts and Orlean did for exotic plants: she attempts to get to the bottom of the mania to touch, to hold, to take, to possess, to hoard, and to control. It's no accident that the subtitles of all three books share the word *obsession*. But if "obsession" has a side that faces the light (passion, enthusiasm, dedication, zeal), it has another side that faces the darkness: acquisitiveness, self-absorption, nihilism, greed.

In their books, Orlean, Childs, and Williams struggle to dissect the urge that pushes some people to set aside "big picture" considerations or ethical constraints and to act in ways that are exclusively self-interested and self-rationalized. That urge is sometimes justified by absence of other economic opportunities—the need to "preserve a way of life," as *Dino Hunters* put it. The amber and mammoth ivory trades are two examples, but so is the trade in blood ivory and in endangered animal pelts and parts generally. Morocco, meanwhile, may be "the world's greatest 'fossil capitalism' economy" (Osborne, 2000), and estimates are that more than 50,000 Moroccans are involved in the mining, sale, and export

there may be few other jobs, though, in 2009, "around a hundred different export outfits batch the bulk of Moroccan fossil and mineral specimens; these are purchased by dealers at the world's major fossil and mineral shows, such as those in Tucson, Arizona, and Munich, Germany. Then they are sold in lots of dozens or hundreds to gift shops in museums and malls" (Sicree, 2009). During a 2018 investigation into the Moroccan fossil trade by the French Magazine, *Libération*, a self-identified fossil "trafficker" explained that

"I locate deposits or traces of bones in books ... and then I go there and dig with the locals." He has already sold hundreds of skeletons around the world, including forty-eight *Basilosaurus* specimens [a predatory whale ancestor that lived in the Eocene] to a single foreign individual. After having excavated them from the ground, [the merchant] takes the ... fossils to his country workshop near Rabat. About ten workers clean them and assemble them into an entire skeleton, even if it means 3D-printing parts or remaking them in plaster. The merchant sells at a local "souk," a market where dozens of exporters flirt with the lack of legal regulation. "I detail the contents of my packages filled with bones to the Ministry of Mines which gives its approval to [the Ministry of] Foreign Trade," he explains, implying that he does not indicate that there is a skeleton inside, whole, ready to be assembled. "All my pieces go with invoices from my Moroccan or American company. It's neither legal nor illegal," he admits. Other traffickers take skeletons out of the country informally, in batches, hidden in piles of stones in cars, counting on getting past poorly trained customs officials.... A ministerial decree of 1994 listed goods subject to export restrictions, such as those of paleontological interest," said Ahmed Benlakhdim, head of the geology department of the Ministry of Mines. As a result, [the Ministry] cannot approve the release of rare specimens, including the bones

BOX only 55 lbs
Brachiopods
150.00—   PS·15

and traces of reptiles, birds and mammals." However, [the fossil merchant] assures [me] that none of his shipments has ever been blocked. (Ollivier, 2018).

THE GREATEST MYTHS EVER TOLD

The conversation about fossil commerce and about commercial and avocational collecting cannot go on as it has in recent decades. Or, rather, it can—provided the parties involved have no real intention of shifting the status quo by so much as a millimeter. As it is, nothing of long-term usefulness will be resolved if the factions do no more than drag their traditional assumptions and conventional world views into the debate and then simply retrench after every skirmish. One thing that keeps getting in the way is a collection of Great Myths:

1. The commercialization of fossils has no effect on professional paleontology or on avocational collectors or, if it does, it is a benign or even helpful one.

No kind of commerce has a completely neutral impact on the system in which it exists; this essay has traced some of the negative effects in the context of paleontology and fossil collecting. There is no question that there are also positive effects. The fact that issues of fossil commerce have become impossible to explore in an "ecological" sense—that is, we cannot objectively assess the collective impact of commercial fossil collecting because attempts to do so are immediately derailed by special-interest bickering—is exactly why the conversation must change.

2. Those with concerns about the commercialization of fossils believe all professional collectors are only out for the money and don't care about science.

Some dealers in fossils work closely with and are trusted by professional paleontologists and are deeply interested in science. Some dealers in fossils are unscrupulous and could not care less about science or ethics. Some fossil dealers are only in it for the money. Some fossil dealers are in it because they've loved fossils all their lives. All of these assertions are true, but none of them need be true of every single individual in every single instance in order to create good-faith policies that benefit everyone.

Meanwhile, constantly pitting one group of facts against the other, as if trying to force a balance scale to tip in one direction or the other, is exhausting and irrelevant. If most fossil dealers are above-board and ethical, that has nothing to do with the reality that black markets, smuggling, and looters and despoilers exist nor does it absolve good actors of responsibility for finding ways to intervene in nefarious practices and processes—precisely because crime and corruption in the fossil world so often take place right alongside legitimate business.

In fact, one of the best ways for fossil merchants and the AAPS to protect "our livelihood" would be to come out strongly in favor of a crack-down on poaching, illegal trade, dishonest dealing, and environmental depredation, to police their own members, and to participate with law enforcement in bringing offenders to justice.

None of the information in this essay proves definitively that fossil dealers are corrupt, greedy science-haters (and it isn't intended to), just as 145 examples of donations to museums by commercial fossil sellers—or 10,145—don't prove they are not. Anecdotes provide context, but, standing alone, they don't prove anything.

3. Critics of the commercialization of fossils are "anti-commercial," "anti-capitalist," or "anti-free trade."

Say they are. That doesn't make their concerns wrong. Sketchy practices in shipping, trading, buying, selling, exporting, and importing of fossils are far from rare. Even when fossil commerce is legal, it has an impact on the planet, on the economic value of and demand for fossils, on casual collectors, and on scientific pursuits. No one is arguing that commercial fossil dealers don't deserve to make a living. They do: just like everyone deserves to make a living. But all kinds of ways of "making a living" are questionable or even illegal. In other words, that's not enough of an argument.

4. Limiting the collection and sale of scientifically important fossils means the end of fossil collecting for everyone.

This is a scare tactic, not an argument.

5. There is no way to craft laws regarding fossil commerce that are fair or balanced because "they" ("academic paleontologists" or "the government") won't allow it.

It is undeniable that a wide variety of positions exist among professional paleontologists regarding commercial collecting, just as it is undeniable that the Society for Vertebrate Paleontology has attempted, for many years, to close fossil collecting off entirely to anyone who is not a "certified" professional. The extreme insistence on that position disenfranchises the entire field of invertebrate paleontology, which is generally overlooked in position-staking over fossil preservation and study, even as it alienates casual and commercial collectors. At the same time, the SVP is legitimately concerned about the egregious abuses it sees and wants to stop them, which is a goal everyone interested in paleontology should share. How do commercial collectors propose to achieve this same goal (other than insisting that abuses don't exist or are "balanced out" by good deeds)?

When words and phrases like "academics," "intellectuals," or "three-piece suits" enter the conversation about fossil collecting and commerce, meanwhile, they are almost invariably dog whistles. For whatever reason, some Americans have let themselves be convinced that "book larnin'" is the enemy of the "common man," a juvenile and dangerous bit of propaganda. In professional and avocational paleontology, such language has come to indicate the distinction between the "cowboys" and "frontiersmen" who are close to the earth from which they are eking out a living, on the one hand, and carpet-bagging, Ivy League, trust-fund snobs on the other. But straw-men aren't getting anybody anywhere.

"Government" is similar. "Government" is code for taxes, regulation, and limitation (bad), while its opposite is "liberty" (good), which, in practice, is nearly always defined solely as a personal right that others are bound to honor, even if threats or violence are required to get them to do so. In the context of fossil collecting, to be fair, "government"—mostly meaning paleontology-related legislation enacted at federal, state, or local levels and applied through the regulations of an alphabet soup of agencies—has frequently shown itself to be a blunt and clumsy instrument. Legislators may typically not know much about paleontology, and so they are swayed by whichever "experts" have their ears. Fossils probably need their own a lobbyist who is neither a scientist nor a collector nor a vendor. Until that happens, the "Seven Kingdoms" approach to legislation and to regulation—whose bannermen form the most powerful army—is singularly incapable of producing equitable solutions.

ASSUMPTIONS OF AUTHORITY

6. Any specimen that is never collected and is lost to the elements or to development justifies all efforts to collect fossils privately off any land and, especially, to profit from them; and

7. The fact that there are unstudied specimens in university and museum vaults means that any attempt at conservation or husbandry of fossil resources is wrong.

Both of these myths are based upon assertions of morality and of "assumptions of authority," as Elizabeth Jones, at the Department of Forestry and Environmental Resources at North Carolina State University, called them in an intriguing analysis of the case of Sue the *T. rex.* Jones traced a decade of litigation and four separate trials all aimed at answering the question "Who owns Sue?" (see Jones' Section 2 for an excellent chronology of the case). In examining the "assumptions" of the interested parties, she wrote:

[The] complex network of individuals involved in Sue's story [ranged] from academic paleontologists and commercial collectors to private citizens, government officials, Native Americans, and the general public. The legal battle over Sue typifies the debate over fossil access in the United States, and the role of scientists and fossil dealers in accessing exceptional fossil finds.... Both individuals and groups staked claim over Sue but justified those claims through different arguments based on different values and assumptions of authority. [Peter] Larson and the [Black Hills Institute of Geological Research, owned by Peter Larson and his brother, Neal], for example, assumed authority over Sue in terms of ownership and expertise. The United States Department of Justice and Department of Interior, along with the Cheyenne River Sioux Tribe, assumed ownership for legal[37] reasons. Scientists, particularly paleontologists of [the Society for Vertebrate Paleontology], assumed authority over Sue on the basis of the specimen's research value, their own scientific expertise, and the need for public accessibility.... Further, regulatory and legislative attempts to control fossil access on government property both before and after Sue's discovery were laden with assumptions of authority regarding the proper relationship between science and capitalism. (Jones, 2019: 9, 24)

In the controversy over revision of the Paleontological Resource Preservation Act (PRPA), first debated under that name in 2001 and finally adopted in 2009, Jones identified additional "assumptions." The 2009 PRPA included a provision that required would-be collectors on public lands to apply and be approved for a collecting permit, though avocational collectors could collect a "reasonable amount" of "common invertebrate and plant paleontological resources," meaning that

[t]he Secretary [of the Interior] ... assumed authority for determining what it meant to be qualified applicant, as well as the meaning of terms like "reasonable amount" and "common invertebrate and plant paleontological resources" [and] the PRPA reinforced scientists' authority in terms of assumed expertise and legal access to fossils on public land. (Jones, 2019: 18-19)

Jones didn't mention, though she might have, the assumption of the AAPS which, along with the American Lands Institute, the American Federation Mineralogical Societies, and many commercial mineral companies, opposed the PRPA (Malmsheimer & Hilfinger, 2003). That is, the assumption that commercial collectors deserved authority over fossils collected on public lands, including the authority to make a profit from them, despite the fact that they are literally public property. There seems something unassailable about maintaining, even if only as a matter of principle, that fossils found on public land should not automatically be made available for private profit, but the

---

[37] For historical reasons related to tribal lands and treaties as well.

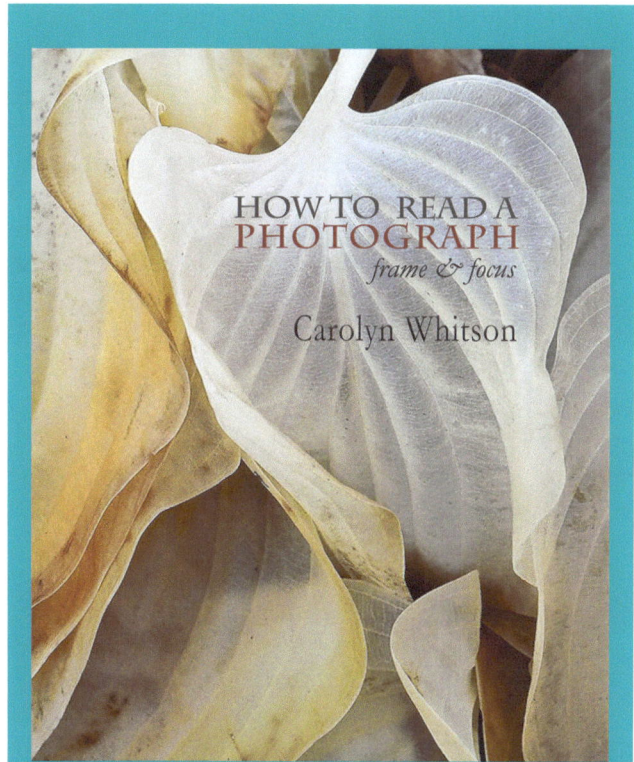

anti-PRPA groups had additional arguments to make. They also feared the PRPA would

> cause "fossils [to] disintegrate shortly after their exposure by erosion" because the bill will "make ... public land[s] a hunting preserve for a privileged few at the expense of permanent loss of most fossils to erosion;" [and] require U.S. museums, which "already have more material than they have staff or monies to house, curate, or exhibit," to hold the paleontological resources collected from public lands. (Malmsheimer & Hilfinger, 2003)

Writing in the related context of institutional archaeological collecting in *Finders Keepers*, Craig Childs confirms similar concerns. "Everything that is removed goes somewhere, he wrote, and

> [m]uch of what has been dug remains undeciphered and unreconstructed. Excavation spoils are piling up. Every major public repository in Arizona will have topped out in five to ten years, a problem faced by institutions across the country and around the world, yet more keeps coming in as cardboard boxes and bags of specimens heap onto each other in storage. Smaller collections across the country can hardly afford to curate what has already been delivered. The U.S. Army Corps of Engineers oversees 50,000 cubic feet of artifacts from field recoveries, and three-quarters of this collection is improperly stored, most of it steadily deteriorating....[38] In the United States, we now have over 200 million individually catalogued objects in the public trust. In addition, there are 2.6 million cubic feet of artifacts stored in bulk whose individual pieces have yet to be catalogued, which comes out to about 1,300 semi truckloads of potsherds, beads, bones, shells, feathers, and buttons. This is what institutional obsession looks like. (2010: 141-142)[39]

These points are repeated in a nearly infinite variety of versions in AAPS literature and by commercial collectors. In "What Commercial Fossil Dealers Contribute to the Science of Paleontology," published in the AAPS's *Journal of Paleontological Sciences*, for instance, the authors wrote that "Any fossil saved from nature is a fossil saved for future generations [because] fossils left in the ground weather away and are lost" (Larson et al., 2014).

The group quoted an even more dire warning by David Martill, a paleobiologist at the University of Portsmouth in the United Kingdom: If fossil collecting were not allowed for "everyone in all countries ... specimens will be lost forever and *the science will die*" (2011; emphasis added).

The thought that uncollected fossils will just "sit there and be destroyed by the elements" or that collected ones will "languish in a museum vault, never to be seen or studied," is painful. At the same time, the reality that museums have more paleontological specimens than they can catalogue or display, or that some fossils are destroyed by natural forces, does not lead to the inescapable conclusion that private individuals have a "right" to ownership of them for the purposes of personal profit (or that researchers have a "right" to remove them to their labs, where they may also sink from view). Nor must we inevitably agree that the solution to the chronic lack of "staff or monies" that makes it difficult for museums to curate their collections is to let fossil dealers sell, for private profit, the fossils that would otherwise be ignored in dusty drawers.

In 2014, Peter Larson and Donna Russell wrote that

> A recent Gallup poll ... shows that 46% of Americans believe that God created humans in their present form. The same poll revealed that 66% of Americans believe that the Earth is less than 10,000 years old. Fifty-four percent believe that creationism should be taught in schools. The challenges that our discipline faces are grave and we need a united front so that we might work together to make fossils more available to the general public, in museums and private collections, so that more people can touch, learn and understand the beautiful story that is the evolutionary history of life on Earth.

It's a noble thought, but fossils in private collections, especially if those collections are overseas—never mind the ones that are transformed into fireplace mantels or paperweights—are infrequently "available to the general public"[40] or for scientific study anyway, nor are they nec-

---

[38] "'The U.S. Army Corps has collections that span the paleontological record,' says Nancy Brighton, a supervisory archaeologist for the Corps. The Corps never set out to amass this prehistoric tome. Rather, the fossils—from trilobites to dinosaurs, and everything in between—came as a kind of byproduct of the Corps' ... large-scale civil engineering projects.... Jen Reardon, an archaeologist with the Corps, says '[M]ost of the Corps's collection ... reside[s] in local museums or universities.... 'The Corps encourages institutions ... to put their specimens on display, but there are just too many to display all at once, she adds" (Imbler, 2019).
[39] In 2008, the Institute of Museum and Library Services reported "that archives, historical societies, libraries, museums, scientific research collections, and archaeological repositories ... hold 4.8 billion artifacts in public trust, of which museums hold 20%" (Manjarrez et al., 2008: 16-17).

[40] In Fall 2019, an intact and unusually well preserved a sabretooth cat skull was accepted for auction by Heritage Auctions (which had also auctioned Prokopi's *Tarbosaurus*) for a minimum asking price of $843,000. The specimen had been found in the tar pits of Los Angeles twenty-two years before the La Brea Tar Pits Museum opened to the public in 1977, and remained unstudied in private collections for more than sixty years. Though the fossil was "considered one of the best examples of a sabre-toothed

essarily "saved for future generations" or for the general benefit.[41] Fossil commerce can also be a way to make fossils "disappear" and does not, in and of itself, necessarily represent an advantage over museum storage or natural destruction.

Meanwhile, commercial collectors who almost literally vacuum up every last tooth, fragment of petrified wood, leaf, bone, or trilobite at a site or are storing tens of thousands of specimens for sale do so not because they want to help the public have access to fossils but because they are safeguarding a commercial inventory. There's no need to judge that behavior as good or bad, but there's no reason to pretend it's synonymous with altruism either.

To put it in other terms, the only way to address these realities is not necessarily to grant authority over public paleontological materials to private commercial fossil dealers who assume they are entitled to advance such a claim because they will do something "useful" with the fossils—namely, turn them into cash.

Here are some other proposals: museums and universities could cull specimens they deem no longer useful for public education or research and sell them to raise money for proper curation efforts and increased staff.[42] Alternatively, they could donate those specimens to private and casual collectors who, in light of the growing difficulty of finding publicly accessible collecting sites, would welcome adding to their collections, or to schools and public-education programs. As a corollary, all sales of fossils by dealers could be taxed a small fraction of a percentage to resolve the problem of under-funded museums and provide them the resources they need to manage their collections properly.

These suggestions are made casually, without any deep consideration of their impact, but they serve to underscore the point that "free market" access to fossils (especially publicly owned fossils) is not the only possible approach. Meanwhile, such possibilities may not be as crazy as they sound. Paleontologist Phil Currie at the University of Alberta in Edmonton, Canada, proposed that the United States create a "philanthropic fund to buy legally collected private specimens and place them in US museums" in order to make sure that scientifically important specimens collected on private land end up in public collections (Pickrell, 2018).

THE END OF EXPERTISE

If it has done anything, the COVID-19 pandemic has made clear the way in which questions of public health and epidemiology can be transformed into political debates over "rights" and "tyranny." The conflict is sometimes viewed as an extension of the so-called "anti-science bias" that seems to have gripped the United States and much of the world in the guise of arguments about climate change, vaccinations, the shape of the planet, the teaching of evolution, and other questions.

To be sure, ignoring information about how disease is transmitted or how physicists determine that the Earth is a sphere can be the result of an anti-science stance, but a more basic reason may be the belief that experts cannot be trusted (because they have "secret agendas") or that they offer nothing more than "opinions" which are no more valid or reliable than those of any other person. More relevant still to the context of professional and avocational paleontology is what Tom Nichols called the "death of expertise" in his 2017 book of the same name:

> I fear we are witnessing the *death of the ideal of expertise* itself, a Google-fueled, Wikipedia-based, blog-sodden collapse of any division between professionals and lay people, students and teachers, knowers and wonderers—in other words, between those of any achievement in an area and those with none at all. (2017: 3)

One of the most common assertions by commercial collectors, echoed in AAPS materials, is that commercial collectors know just as much as "professional" or "academic" scientists. The vexing term "commercial paleontologist" has taken hold, though people who sell specialized equipment for orthodontists don't call

---

cat skull in existence" (Folven, 2019), bids failed to meet the minimum price and, as of July 2020, it was still listed as "not sold" in Heritage's online catalog and remains unavailable to the public. Dr. Thomas Carr's 2015 "*T. rex* List of Shame" (http://tyrannosauroideacentral.blogspot.com/2015/02/tyrannoethics-4-naturalis-t-rex-t-rex.html) includes fourteen American tyrannosaur specimens that were sold into the hands of private parties and, thus, essentially lost to science. A fifteenth, "TAD," was privately purchased by a foundation in Hong Kong in 2018 and first displayed there at a shopping mall.

[41] Paleontologist Dr. David Hone wrote in 2012: "Scientists make a policy of only studying specimens held in public (generally state owned and run) museums. There are some exceptions, but in general if it's not there, you can't work on it, or at least journals won't let you publish on it. That naturally gives us a lot to work with, but good, even great, specimens are in private collections and private museums. True, [private collectors and museums] do sometimes hand over their material ... but there's no guarantee they ever will.... And therein lies the problem.... In short, research is losing out massively to collectors."

[42] This is complicated. Museum collections often yield surprising discoveries when "forgotten" material is reexamined in light of new scientific information. In 2015, for example, paleontologists Dean Lomax and Judy Massare discovered a new species of ichthyosaur in the collections of the Doncaster Museum and Art Gallery in South Yorkshire, England. The specimen was believed to be a plaster copy and had been unexamined for more than thirty years (Gil, 2015). Lightning struck a second time in 2020 when researchers recognized yet another new ichthyosaur, collected decades earlier, in storage at the Museum of Natural History in Stuttgart, Germany ("New Species," 2020).

themselves "commercial dentists," and professional landscapers or nursery keepers don't typically demand to be known as "commercial botanists." As a kind of reverse discourse intended to disarm disrespect and to emphasize commitment, experience, and skill, it's an understandable strategy. But it still causes, as Nichols put it, the collapse of a distinction that doesn't really deserve to fall. Yes, someone who has been digging trilobites out of a quarry in Michigan for twenty years certainly knows more about the species of trilobites found there, about how they are preserved in that exposure, and about the fauna and flora typically associated with them than does a vertebrate paleontologist whose career has been spent on Jurassic corals. But that doesn't make him a paleontologist.

And yet this refusal to acknowledge one kind of expertise—in order to insist upon another—is a common tactic in the fossil-dealer wars. Mike Triebold told *Wired* in 2018, "There are some people in academia that think that if you're not a degreed paleontologist, you're not qualified to work on [dinosaurs]. That's absolute hogwash. It's bullshit" (Reynolds, 2018). Five years earlier, Triebold told *High Country News* that he "believes data collection by commercial operations can be superior to that done by academic institutions strapped by budgets and bureaucracy" (Hodges, 2013).

In 2014, Kenshu Shimada (DePaul University and Sternberg Museum of Natural History), Philip Currie (University of Alberta), Eric Scott (Division of Geological Sciences, San Bernardino County Museum), and Stuart Sumida (California State University, San Bernardino) published an article entitled "The Greatest Challenge to 21st Century Paleontology: When Commercialization of Fossils Threatens the Science" in *Palaeontologia Electronica*. Peter Larson of the Black Hills Institute of Geological Research and Donna Russell of Geological Enterprises responded a month later in the same journal, and Neal Larson, at the time the President of AAPS, along with Walter Stein, the owner of PaleoAdventures; Michael Triebold, the president of Triebold Paleontology; and George Winters, AAPS's Administrative Director, developed the response into a longer essay that appeared in the AAPS's *Journal of Paleontological Sciences* in October 2014.

Neal Larson's group wrote:

> There is a common misconception among many academic paleontologists ... (Shimada et al. included) that commercial collectors are not professionals, nor true paleontologists....[43] As independent commercial paleontologists and members of the Association of Applied Paleontological Sciences (AAPS), many of us make our living in paleontology mostly in

collection, preparation and display. While only a few independent paleontologists have a degreed education in the field, many have extensive training, especially in collection and preparation of fossils.... Some independent paleontologists have published, working both alone and in conjunction with academic paleontologists to report discoveries, describe new species, and publish on the taxonomy, taphonomy, biostratigraphy, biology and geology of the fossils and rocks we work with.

Peter Larson and Kristin Donnan (2014), meanwhile, writing in their book, *Rex Appeal*, Larson's version of the saga of Sue the *T. rex* and his conviction on charges of fossil trafficking and customs violations unrelated to Sue, argued that "the demonization and marginalization of a specific portion of the paleontological community is the result of misunderstanding, *misplaced entitlement* and simple intolerance" (2014; emphasis added), and they went on to say that "legitimate independent collectors ... are as trained, careful and thorough as academic crews— often more so" (359), adding that "collectors' affiliations or degrees should not matter; their knowledge, experience and how much they care should" (349) (the word "legitimate" is not defined).

Lamenting the refusal of the Society for Vertebrate Paleontology to adopt the recommendations of the 1987 National Academy of Sciences' "pro-collecting" report, which Peter Larson helped write, the authors explained that

> an SVP committee circulated a position statement warning against "documented cases of unethical practices by commercial collectors, commercial operations masquerading as legitimate academic institutions, the great increase in sales of scientifically important specimens to interior decorators and hobbyists, and the rapidly inflating prices that create profound security problems for institutional collections. I [Peter Larson] know the SVP committee did not represent the entire membership in its position; *only a small, vocal group wore three-piece suits*. (103-104; emphasis added)

Paleontologist Robert Bakker, who contributed the Introduction to *Rex Appeal*, compared Peter Larson and the personnel of the Black Hill Institute to

> [t]he Sternberg family of Kansas ... three generations of independent fossil hunters who were held to be saints by my professors at Yale in the 1960s. The Sternbergs eschewed the cap-and-gown trappings of university training.... But there was another element in the community of fossil scholars, a generation of Ph.D.s who had a caste mentality, a belief that only university scholars were entitled to dig precious fossils. (xi-xii)

---

[43] For the sake of accuracy, Shimada and his colleagues neither wrote nor implied anything of the kind.

This way of marking out perceived power differences is a recognizable shorthand for what are most commonly imagined as class distinctions: the working man vs. the pampered, soft-palmed egghead, the blue collar rather than the white. But it's curious in the context of what is simultaneously a challenge to and an assertion of expertise, and Bakker's comments about "cap-and-gown trappings" are downright puzzling from a man with degrees from both Yale and Harvard.

A similar skepticism about expertise—at least the kinds of expertise that come from "academia"—pervades *Dino Hunters* in which, according to the Discovery Channel's promo, "Montana rancher Clayton Phipps and Wyoming fossil hunter Mike Harris buck the academic status quo." On *Dino Hunters*, Phipps proudly tells the camera, "I don't do the science. I'm just a cowboy who likes fossils," and, in 2013, he told a reporter:

The (professional) bone hunters don't have a really good reputation around here, because a lot of the academics haven't done much for our community. They've come in and said, "Oh, yeah, we are going to study this," then no one hears anything about it after they leave. (Hodges, 2013)

It isn't clear how many paleontology or paleobiology journals Clayton Phipps reads on a regular basis, but that's where one would "hear about" studies once they were completed. Nor does *Dino Hunters* reveal how Phipps and his company "benefit the community" in ways that are more beneficial than what "academics" fail to do. "Bucking the academic status quo," finally, probably sounds like a reasonable pastime to someone who wants others to see him as a "cowboy" or whose goal is to own the elites, but what exactly is the "academic status quo" the dino hunters are "bucking"? Formal education; painstaking research that sometimes takes years, a stock of information that is both broad and deep within an area of specialization; interactions with peers doing similar work all over the world, and assertions backed by data and revised over time? If so, in 1871 Darwin said everything that needed saying about such attitudes: "ignorance more frequently begets confidence than does knowledge."

Skirmishes such as these, in any case, typically take place on the perceived borderlands between professions and spheres of expertise, and Elizabeth Jones noted that fossils themselves were one of the foci of "boundary work" in paleontology—that is, the "process ... by which individuals and groups constructed or deconstructed boundaries concerning fossil access, then implemented and enforced those boundaries in order to establish authority" (4). Fossils can be shared and appreciated across boundaries (by scientists, avocational collectors, and vendors, in this case), but their meaning and importance are vastly different to each group.

Law professor Alexa Chew argued that

[t]here are at least three ways to assign value to a fossil: economic, scientific, and cultural. Preserving one type of value may limit or destroy the others. For example, digging up a fossil as quickly as possible to rush it to market may maximize economic value, but it can destroy scientific value. For private collectors, the value of a fossil is inherent in the object itself; its value is akin to that of a piece of art.... For scientists, however, the value of a fossil is usually contextual. Most paleontological research focuses not only on the fossil itself [and] .... [w]ithout contextual information, evidence of how extinct organisms lived is lost.... Finally, a fossil's cultural value reflects its importance to the larger community. The value of certain objects transcends market value or scientific discovery to become part of collective knowledge and experience. Such objects, including fossils, intuitively belong within the public domain" (Chew, 2005: 1033-1034)[44]

No different—and no less subject to conflicts and hostile interpretations—are the ways in which each area "develop[s] ... and enforce[es] a clear set of methods to 'discipline' the information [that comes from fossils]" (Jones, 2020: 10).

Quoting Thomas Gieryn, professor emeritus of sociology at Indiana University, Jones expanded on the idea, noting that

boundary-work is a process by which scientists construct, deconstruct, and negotiate definitions of what (in their view) gets to count as science. But as [Gieryn] argued, there is no one way to do science ... [and] 'boundary-work becomes a means of social

[44] There is one other way to assign value to fossils: for what they supposedly reveal about the Biblical flood and the age of the planet. Various groups of creationists and "Young Earthers" are among the most assiduous of fossil collectors (and buyers) and are a presence at many dig sites across the country. In 2002, an *Allosaurus* skull and "the claws to a 100-foot Sauropod, presently believed to be of the rare *Ultrasaurus* variety" were discovered in Colorado, though many details (such as who discovered the fossils and when) are disputed. As so often happens with finds of outstanding dinosaur fossils that can be sold for high-dollar amounts, this specimen ended up in litigation. When all was said and done, the specimen was donated to Ken Ham's Creation Museum in Petersburg, Kentucky, where it "is now one of the museum's main attractions." After the donation, the museum reappraised the skeleton (now acknowledged as *Allosaurus*), and "declared its value at a million dollars.... 'The intact skeleton of this allosaur is a testimony to a catastrophic, rapid burial, which is confirmation of the global Flood a few thousand years ago as recorded in the Bible,' the Creation Museum says on its web site'" (Bethea, 2019).

control.' … In particular, scientists build boundaries when they feel their authority is at stake. In many cases, scientists are seeking to defend their interests by demarcating their work from other interests they consider to be non-scientific at best or detrimental to science at worst. (21)

"In Fossil Wars," Paige Williams, author of *The Dinosaur Artist*, summarized the "stark opposing arguments" this way:

Commerce: overregulation destroys the public's interest in the natural world.

Science: commodification compromises our evolving understanding of the planet.

Commerce: science doesn't need hundreds or even dozens of specimens of one species.

Science: multiple specimens elucidate an organism and its environment over time.

Commerce: private collectors wind up donating their stuff to museums anyway.

Science: specimens collected under nonscientific conditions are worthless to research.

Commerce: most museum fossils land in storage, never to be studied.

Science: stored fossils have generated profound advances decades after their discovery.

Commerce: scientists are stingy and elitist, with their snooty PhDs.

Science: commercial hunters are destructive and greedy. (Williams, 2018b)

There's no point denying the existence of professional paleontologists who, like scientists in allied fields, look with disdain on unaccredited collectors and commercial dealers, nor that they defend their perceived "territories."

What becomes blatantly obvious in analyzing the "fossil-commerce wars," though, is that commercial collectors, particularly in the formal forum offered by a trade organization, are engaged in precisely the same process: constructing and policing boundaries to separate their work from actions they consider to be anti-commerce, unconstitutional, elitist, unscientific, or detrimental to their economic interests.[45]

---

[45] If "elitism" in the paleosciences is objectionable, there's no better example than David Martill, on whom Larson et al. (2014) relied in part and of whose views they presumably approve. In a 2011 article in the Geological Society of London's *Geoscientist*, Martill contrasted what he called the "obscene" destruction of fossils by mining companies in Brazil with the "draconian" laws that outlawed private fossil collecting and strictly regulated scientific work. In Martill's view, it was "endemic," "grass-root corruption" that created the global trade in Brazilian fossils and led "poor farm workers … to dig deep and dangerous excavations." In other words, Martill made and defended a moral choice and

To put the finest point possible on these matters, Larson and Donnan included an appendix in *Rex Appeal*: "The Modern Bone War." In it, they present a table that separates what they considered the "establishment view" of commercial fossil collecting from the "independent view"—a prolonged exercise in boundary work. A good bit of what they characterize as the "establishment view" is a series of straw men, and much of the "independent view" is either personal opinion asserted without evidence or fails to respond logically to the positions the "establishment view" is accused of holding. But the "independent view" remains an argument for the expertise of non-scientists and for their "authority" over "establishment" expertise where points of friction exist.

WHAT IS TO BE DONE?

Selling fossils is not per se a moral issue—and the AAPS would likely agree. But commercial practices can be immoral, unethical, or illegal. Virtually no one would argue that selling pizza is a moral question, but deliberately selling pizza with cheese known to be contaminated is immoral, and selling pizza stolen from a competitor is both unethical and illegal. But none of those issues turns the selling of pizza, in and of itself, into a moral quandary. So let's stop pretending that's the issue with fossils. The questions should be, instead, what are the broader consequences of selling fossils? What limitations on fossil commerce can exist side-by-side with reasonable opportunities for casual collecting, scientific study, and ethical bartering, selling, and buying? What impact do market demand, collecting, and the potential for over-collecting have on fossils as a natural resource; on other human beings; and on the ability to increase knowledge and to preserve and disseminate information about natural heritage?

At the same time, fossil collecting is not a "right." It is a claim on a natural resource, and it can be managed in substantially the same way as are other, similar claims. The majority of the individuals who assert "rights" in these matters, meanwhile, are men of a remarkably similar ethos, and it cannot continue to escape us how often the issues of possession, sovereignty, entitlement, and territory (that is, figurative and literal power) underlie what are disguised as more "secular" conflicts over science vs. commerce, professionals vs. amateurs, enforcement vs. cooperation, institutional authority vs. personal passion.

The reality is that none of us owns or controls anything for very long—something on the order of six or seven decades, if we're lucky. Particularly in the realm of fossils—which existed tens and even hundreds of millions

---

disguised it as scientific objectivity: In pitting dangerous work conditions, systemic poverty, and corruption against science, science naturally deserved to triumph. That's the definition of academic elitism.

of years before human beings and will presumably exist long after we're gone—the singular focus on ownership, control, and "rights" seems both absurd and petty.

Writer and Graduate Director of World Cultures and Literature at the University of Houston, Alessandro Carrera, addressed these considerations in a 2020 essay. "The snake in the Garden of Eden is that terrible little word, 'rights,'" he wrote.

> [R]ights exist as such before the law, not before the community. With regard to the community, an individual has only obligations, and the community has obligations to him or her.... The confusion between rights and obligations always destroys communities.... Along with puritanism and the claim to an "illuminist" personal enlightenment inaccessible to the rest of humankind, extreme libertarianism—meaning the total rejection of the idea of civil society (one based on obligations and not on rights)—has played a part in making American what it has been and what it is today. (Carrera, 2020)

In such a context, the entire debate about fossil collecting, sales, and ownership becomes distorted, and resolution is hopeless as long as the crux of the matter must remain whether scientists have the right, or avocational collectors have the right, or commercial fossil dealers have the right. The more crucial questions are: On what basis do any of them assert a right (or, more philosophically, on what basis do they assert a right to assert a right?)? What obligations do they recognize to others interested in the same "goods," and how do they plan to honor them?

The commercial fossil dealer has a duty to respond to the concerns of the scientist and the avocational collector, whose interests are no less important because they do not involve commerce or real estate. The avocational collector has a duty to scientists and to commercial collectors, who often make casual collecting possible. The scientist has a duty to the private collector and to the person who makes a living from fossil sales, both of whom are frequently allies and partners. And all three have a duty to their communities and to the Earth in the broadest sense of the word: its ancient and recent past, its integrity and health, its fragility and uniqueness, its human and non-human inhabitants, and its future, which stretches far beyond the lifetimes of any individual fossil dealer, amateur collector, or paleontologist reading these words.

Fossils are a finite, non-renewable natural resource. That is simply a fact. Fossils are subject to erosion, the elements, and damage from land development and industry, but they are also vulnerable to over-collection, vandalism, and commercial exploitation. They deserve oversight. (Yes, some fossils are extremely abundant, but that doesn't make them either renewable or infinite.) U.S. and international laws and treaties recognize the importance of reducing harm to the natural world (deforesta-tion, water quality, overfishing, habitat destruction) as well as of monitoring the exploitation of plants, marine and other mollusks, seahorses, ornamental fish, exotic or endangered animals, and other species.

In fact, since at least the 1980s, legislation and regulation at both the single-country and international level have been adopted to address the harm to organisms that are affected by over-harvesting for commercial sale. There is essentially no disagreement that the commercial trade in seashells has reduced both populations and catch rates, for example, or that it is driving some species to extinction (see, e.g., Deines, 2018). Commercial shelling is closely analogous to commercial fossil collecting, and the same objections to regulation are raised by dealers and exporters. But if governments have so widely recognized that regulation is necessary to protect literally thousands of species of animals and plants, the need for similar oversight regarding fossils seems self-explanatory. Martill (2011) and his adherents insist that fossils don't need protection because they're dead, but that's a sophist's argument. It's precisely because they are dead—and because they cannot "bounce back" if left alone—that regulation is important.

At the same time, "all fossils are rare" and "fossils aren't at all rare" are both falsehoods. When asserted, they are expressions of personal, non-empirical positions about nature, and they are too vague to be taken seriously. Some fossils are abundant; some fossils are not. A strict preservation ethic is inappropriate for common fossils, even if it is appropriate for rare ones.[46] Treating common fossils as

---

[46] Dr. David Hone, the paleontologist, science blogger, and author of *The Tyrannosaur Chronicles*, assessed the situation this way: "[W]ith willing buyers out there [and] many more dealers and collectors looking for fossils than the researchers ... we tend not to find the best stuff, and we can't afford to (and in many cases morally shouldn't) buy them. In short, research is losing out massively to collectors. The problem though is a very complex one. Good fossil dealers ... are quite happy to hand over, or sell at a discount, or at least give first refusal to museums for good and important specimens. Generous owners do donate their collections or individuals specimens to museums. Many people will develop an interest in palaeontology ... they might not have had otherwise from the purchase or gift of some small trilobite or shark tooth and I don't think any palaeontologist would begrudge them it. The problem lies in where this kind of dealing should stop. Selling an ammonite of a species represented by thousands or even millions of specimens? Sure, go for it. How about some new and incredible species of dinosaur, preserved with a dead mammal in [its] stomach, a set of eggs in [its] body and ... skin and feathers? Absolutely not. What about a dinosaur foot though? Or half a skull? Or a single caudal? Or a broken tooth? Every specimen can add some information ... but there is understandably a huge grey area ... that is only compounded by confusing laws and regulations.... Where does this leave us? In a mess frankly, but ... the lack of clear national and international regulations and the lack of enforcement means that valuable specimens are being lost to science."

if they were rare impedes research and harms avocational, non-commercial collecting. Treating rare fossils as if they were common excuses haphazard collection methods, robs science of opportunities to increase knowledge, tends to reduce access to fossils to those with means, and provides a handy excuse for sidestepping the issue of (all) collectors' responsibility to the public trust.

In a similar vein, asserting that commercial collecting has zero impact on fossil resources, on the environment, and on other living things, including human beings, is hopelessly (or willfully) naïve. To persist in that conviction is to signal the belief that private advantages and personal interests are inherently more valid than the concerns of a wider community, that opinion deserves more serious consideration than expertise; and that stubborn refusal to acknowledge the effects of actions on other individuals or groups is more "principled" than thoughtful reflection or a sense of general regard—in other words, that rights are more meaningful than obligations. Each and every one of these is a moral choice, not a business decision or a political position.

A more promising approach may be to think of fossils as a "commons." That doesn't mean everyone owns everything or that the "state" owns everything—or any of the other nightmare scenarios that such a word may evoke among those convinced that "socialists" lurk behind every bush. Rather, it means that certain cultural and natural resources are shared, and that groups of individuals (communities, groups of users) manage those resources for individual and collective benefit. Every brachiopod pulled from a southwestern Ohio roadcut probably doesn't need managing, but some fossils—and access to some fossils—clearly do.

For such a fossil-resource-management strategy to work, all the principals will have to abandon the view that the world was made for them to own, to control, and to dispose of as they personally see fit. They'll have to move beyond the narcissistic certainty of their own exceptionalism—bad actors may exist, but they are good; others may be motivated by base impulses, but theirs are pure; others may have a claim on goods and resources, but theirs are more valid. Fantasies of personal dominion over the earth must be surrendered, and "don't tread on me" has to be replaced by an understanding of duties to other people, the community, the planet, and the future.

This is not a battle that must be won or a point that must be proven. It is a consensus that must be reached. Pious calls to "work together" are pointless if what they mean is that others should work to come around to your position.

Issues such as these may seem a long way from the question of whether to sell fossils or own fossils or conserve fossils as a national and scientific legacy, but they aren't: they are the absolute heart of the matter.

HOW NOW, FOSSIL TRADE?

The elements that follow are essential to the code of ethics of any professional or trade organization that intends to represent commercial fossil collectors, propose or oppose laws, or otherwise advocate on behalf of commercial collectors. They're no less essential to casual collectors or to avocational fossil-and-mineral groups because collectors of every kind could benefit from a broader concept of ethics and responsibility in the context of paleontological resources.

- Recognize that fossils are a finite and limited natural resource, and that only a long-term ecological approach to their management can serve as the foundation for discussions regarding law and policy. Failing to be concerned about the natural environment in the 21st century is like failing to be concerned about the Black Death in Europe in the 14th, and a conversation that does not acknowledge fossils as an natural resource—and not simply as a product to market, a treasure to hoard, or a statistical data point—is a useless one.

- Acknowledge that the impact of commercial collecting cannot be neutral, no more than it can be solely beneficial or solely detrimental. It cross-fertilizes and is cross-fertilized by other realities. Only in that wider framework can dealers, collectors, and scientists assume responsibility for the broad, interconnected impact of the fossil market, and only then can professional societies, trade associations, accredited paleontologists, business owners, and civilian scientists craft equitable, sustainable approaches to fossil collecting and commerce.

- Adopt and enforce a strict, meaningful provenance requirement. Oblige members to possess proof of provenance for every fossil or mineral they sell or barter (and to display or provide that proof to buyers) with a value above a limit to be established.[47]

- Adopt a regulation that clearly prohibits the false labeling of paleontological materials sold by members, including whether specimens have been restored, reconstructed, repaired, or composited. Establish an enforcement mecha-

---

[47] "Paleontologists ... are pleading with private collectors to demand proof of a fossil's origins before they buy—just as they would question the pedigree of a painting or an antique.... I'm saying, 'Ask for provenance,' [Dr. Mark A. Norell, Chairman of Paleontology at the American Museum of Natural History] said. 'It worked in the art world, but it hasn't hit the fossil world.'"

nism that includes, where applicable, the automatic reporting of violations to law enforcement.

- Introduce a paid permitting system for commercial collecting and selling, with funds earmarked for public natural history museums, and work for its adoption nationally,

- As a question of professional ethics, a member who works for or runs a museum should not also be a dealer in the things curated in the museum.

- Deny membership to anyone who has been convicted of any crime related to fossil commerce.

- As an organization, join and support boycotts of, prohibitions against, and oversight regarding merchandise like Burmese amber and ivory rather than grudgingly acknowledge protection efforts as "noble causes" while still implying that their most important impact is harm to business.

- Release statements, write articles, send emails to members, and post on social media whenever a case of smuggling, theft, inappropriate collecting practices, or bad dealer behavior comes to light and educate the public about why the behavior is unacceptable.

- Voluntarily cooperate with any government agency in investigations of smuggling, poaching, or other misconduct.

- Prohibit members from participating in trade shows or fairs in which other dealers do not comply with these guidelines.

To piggyback onto a phrase that has been used recently in a different context: I honestly don't know how to convince the leadership of AAPS or other commercial or avocational fossil collectors that they should care about the welfare of the natural world, about endangered species, about other people, about the loss to science of important fossils, and about the risks of unregulated commerce. All I know is that *not* caring is a choice, and the implications of that choice are staggering.

WORKS CITED

Abel, H. (1996, 4 March). Who Owns These Bones? *High Country News.* Retrieved from hcn.org/issues/54/1673.

Actman, J. (2016). Woolly Mammoth Ivory Is Legal, and That's a Problem for Elephants. *National Geographic.* Retrieved from nationalgeographic.com/news/2016/08/wildlife-woolly-mammoth-ivory-trade-legal-china-african-elephant-poaching.

Adams, W. (2018, 2 July). Who Owns Archaeological Artifacts? *Relic Record.* Retrieved from relicrecord.com/blog/who-owns-archaeological-artifacts.

Association of Applied Paleontological Sciences (2014, 6 October/2020, 8 June). Fossil Specimens Placed in Museums and Universities by Commercial Paleontology. *The Journal of Paleontological Sciences.* Retrieved from aaps-journal.org/Commercial-Contributions-to-Paleontology.html. Update of 8 June 2020.

Austin, E. (2020, 29 January). Sharks in Kentucky? What explorers found in Mammoth Cave is blowing researchers' minds. *Louisville Courier Journal.* Retrieved from courier-journal.com/story/news/local/2020/01/29/mammoth-cave-sharks-rare-fossils-found-kentucky-national-park/4565713002.

Barrett, P. M. and Johanson, Z. (2020). Editorial: Myanmar (Burmese) Amber Statement. *Journal of Systematic Palaeontology.* Retrieved from 10.1080/14772019.2020.1764313.

Bauwens, J. (2020, 2 July). Landslide at Myanmar Jade Mine Kills Over 100. *SciencyThoughts.blogspot.com.* Retrieved from sciencythoughts.blogspot.com/2020/07/landslide-at-myanmar-jade-mine-kills.html.

Ben Yahia, J. (2019, 7 June). Morocco's Surging Trade in Fossils. *Enhancing Africa's Ability to Counter Transnational Crime (ENACT) Africa.* Retrieved from enactafrica.org/research/trend-reports/moroccos-surging-trade-in-fossils.

Bending, Z. (2019, 28 August). Why We Need to Protect the Extinct Woolly Mammoth. *TheConversation.com.* Retrieved from theconversation.com/why-we-need-to-protect-the-extinct-woolly-mammoth-122256.

Bethea, C. (2019, 1 October). Mark Meadows and the Undisclosed Dinosaur Property. *The New Yorker.* Retrieved from newyorker.com/news/news-desk/mark-meadows-and-the-undisclosed-dinosaur-property.

Bradley, Lawrence W. *Dinosaurs and Indians: Paleontology Resource Dispossession from Sioux Lands.* Denver, CO: Outskirts Press.

Breznican, A. (1997, 17 February). Pistol Packing Paleontologist: Pitt Grad Fights Crime to Save Precious Fossils. *The Pitt News, 91*(88), 1, 3.

Brown, K. (2019, 21 February). Forget the Old Masters, It's All About the Old Monsters in the Booming Market for Dinosaur Fossils. *artnet News.* Retrieved from news.artnet.com/market/dinosaur-fossil-market-1454464.

Browne, M. L. (1994, 15 February). Clash on Fossil Sales Shadows a Trade Fair. *New York Times,* C1, C9.

Browne, M. L. (1997, 5 October). Tyrannosaur Skeleton Is Sold To a Museum for $8.36 Million. *New York Times,* 37.

Cabrera, J. (2017, 3 March). Morocco Blocks Sale of Dinosaur Skeleton. *Morocco World News.* Retrieved from moroccoworldnews.com/2017/03/209942/morocco-blocks-sale-dinosaur-skeleton.

Carrera, A. (2020, 19 July). COVID e la Fine del Sogno Americano. *DoppioZero.com.* Retrieved from doppiozero.com/materiali/covid-e-la-fine-del-sogno-americano.

Chapple, A. (2016). The Mammoth Pirates: In Russia's Arctic North, a New Kind of Gold Rush is Underway. *RadioFree Europe/Radio Liberty.* Retrieved from rferl.org/a/the-mammoth-pirates/27939865.html.

Chew, A. Z. (2005). Nothing besides Remains: Preserving the Scientific and Cultural Value of Paleontological Resources in the United States. *Duke Law Journal, 54*(4), 1031-1060.

Childs, C. (2010). *Finders Keepers: A Tale of Archaeological Plunder and Obsession.* New York: Little, Brown, and Company.

Coates, J. (1991, 10 November). Rustlers Finding Gold in Dinosaur Bones. *Chicago Tribune.* Retrieved from chicagotribune.com/news/ct-xpm-1991-11-10-9104110122-story.html.

Collins, J. (2019, 23 August). CITES: Elephant Ivory Ban Upheld, but Legal Loopholes Remain. *Deutsche Welle/DW.com.* Retrieved from dw.com/en/cites-elephant-ivory-ban-upheld-but-legal-loopholes-remain/a-50118236.

Colorado Historical Society (n.d.). Strategies for Protecting Archaeological Sites on Private Lands. Office of Archeology and Historic Preservation Publication #1617. Retrieved from historycolorado.org/sites/default/files/media/documents/2019/1617.pdf.

Committee on Guidelines for Paleontological Collecting (1987). *Paleontological Collecting.* National Research Council Board on Earth Sciences, Commission on Physical Sciences, Mathematics, and Resources. Washington, DC: National Academy Press.

Conniff, R. (2019, October). Inside the Homes (and Minds) of Fossil Collectors. *National Geographic,*

Dalton, R. (2007, 25 October). Laws under Review for Fossils on Native Land. *Nature, 449,* 952-953.

Deines, T. (2018, 16 July). Seashell Souvenirs Are Killing Protected Marine Life. *National Geographic.* Retrieved from nationalgeographic.com/animals/2018/07/wildlife-watch-seashells-illegal-trade-handicrafts.

Denig, Edwin Thompson (1961). *Five Indian Tribes of the Upper Missouri: Sioux, Arickaras, Assiniboines, Crees, Crows.* John C. Ewers, Ed. Norman, OK: University of Oklahoma Press.

Dinosaur Hunter Sentenced for Stealing Fossils (2009, 24 June). Science on NBCNews.com. Retrieved from nbcnews.com/id/31533733/ns/technology_and_science-science/t/dinosaur-hunter-sentenced-stealing-fossils/#.XxeIN21Khow.

Dinosaurs & Natural History (2019). *Aguttes.com.* Retrieved from expertise.aguttes.com/en/dinosaurs-natural-history.

Displaced by Clashes over Amber, Villagers in Myanmar's Kachin State Risk Death to Secretly Mine It (2019, 7 November). *The Irrawaddy.* Retrieved from https://www.irrawaddy.com/specials/displaced-clashes-amber-villagers-myanmars-kachin-state-risk-death-secretly-mine.html.

Dyke, G. (2015, 65 January). Fossil Collecting Should be for Everyone—Not Just Academics. *The Conversation.* Retrieved from theconversation.com/fossil-collecting-should-be-for-everyone-not-just-academics-34830.

Espinoza, Edgard O. and Mann, Mary-Jacque (1993). The History and Significance of the Schreger Pattern in Proboscidean Ivory Characterization. *Journal of the American Institute for Conservation, 32*(3), 241-248.

Espinoza, Edgard O. and Mann, Mary-Jacque, with the cooperation of the CITES Secretariat (2010, 27 January). Identification Guide for Ivory and Ivory Substitutes. Ashland, OR: U.S. Fish and Wildlife Service Forensics Laboratory. Retrieved from fws.gov/lab/ivory.php.

Finkel, E. (2019). This Australian Farmer Is Saving Fossils of Some of the Planet's Weirdest, Most Ancient Creatures. *Science, 363*(6434), 1382-1385.

Flanagan, J. (2019, 22 August). Ivory Smugglers Use Mammoths to Pull Wool over Eyes of Customs. *The Times* (London): 36.

Folven, E. (2019, 26 September). Auction Digs Up Rare Fossilized Saber-Toothed Cat Skull. *Beverly Press, 29*(39), 1, 30. Retrieved from beverlypress.com/wp-content/uploads/2019/09/9.26-issue.pdf.

Gammon, K. (2019, 2 August). The Human Cost of Amber. *The Atlantic.* Retrieved from theatlantic.com/science/archive/2019/08/amber-fossil-supply-chain-has-dark-human-cost/594601.

Gettleman, J. (2012, 3 September). Elephants Dying in Epic Frenzy as Ivory Fuels Wars and Profits. *New York Times.* Retrieved from nytimes.com/2012/09/04/world/africa/africas-elephants-are-being-slaughtered-in-poaching-frenzy.html

Gil, Victoria (2015, 19 February). Forgotten Fossil Found to be New Species of Ichthyosaur. *BBC News.* Retrieved from bbc.com/news/science-environment-31521719.

Graslie, E. and Gimbel, A. (2020). Examining the History of Fossil Collection from Native American Lands (2020). *Prehistoric Road Trip.* Chicago, IL: WTTW National Productions. Retrieved from interactive.wttw.com/prehistoric-road-trip/stops/examining-the-history-of-fossil-collection-from-native-american-lands.

Grimaldi, D. (2019, 8 June). Supporting a Boycott of Burmese Blood Amber. *New Scientist,* 3233, 27. Retrieved from newscientist.com/letter/mg24232330-100-supporting-a-boycott-of-burmese-blood-amber.

Helmore, Edward (2019, 24 February). Dinosaur Fossil Collectors "Price Museums Out of the Market." *The Guardian.* Retrieved from theguardian.com/science/2019/feb/24/dinosaur-fossils-collectors-museums-price-sale.

Helmore, Edward (24 February 2019). Dinosaur Fossil Collectors "Price Museums out of the Market." *TheGuardian.com.* Retrieved from theguardian.com/science/2019/feb/24/dinosaur-fossils-collectors-museums-price-sale.

Hodges, M. (2013, 26 August). Dinosaur Wars. *High Country News.* Retrieved from hcn.org/issues/45.14/dinosaur-wars/print_view.

Hone, D. (2012, 4 February). Fossil Collecting – A Delicate Balance Act. *Dave Hone's Archosaur Musings.* Retrieved from archosaurmusings.wordpress.com/2012/04/02/fossil-collecting-a-delicate-balance-act.

Hopkin, Michael (2007, 17 January). Palaeontology Journal Will "Fuel Black Market." *Nature, 445,* 234-235. Retrieved from doi.org/10.1038/445234b.

Imbler, S.(2019, 7 August). Why Does the U.S. Army Own So Many Fossils? *Atlas Oscura.* Retrieved from atlasobscura.com/articles/why-does-the-army-own-dinosaurs.

Ingraham, C. (2017, 21 December). American Land Barons: 100 Wealthy Families Now Own Nearly As Much Land As That of New England. *Washington Post.* Retrieved from washingtonpost.com/news/wonk/wp/2017/12/21/american-land-barons-100-wealthy-families-now-own-nearly-as-much-land-as-that-of-new-england.

Joel, L. (2020, 11 March). Some Paleontologists Seek Halt to Myanmar Amber Fossil Research. *New York Times.* Retrieved from nytimes.com/2020/03/11/science/amber-myanmar-paleontologists.html.

Jones, Elizabeth (2019, 31 December). Assumptions of Authority: The Story of Sue the *T. rex* and Controversy over Access to Fossils. *History and Philosophy of the Life Sciences, 42*(2), 1-27. Retrieved from doi.org/10.1007/s40656-019-0288-4.

Kertscher, T. (2019, 10 September). Evidence Shows Most of the 47 Men in Famous 'Declaration of Independence' Painting Were Slaveholders. *PolitiFact.com.* Retrieved from politifact.com/factchecks/2019/sep/10/arlen-parsa/evidence-shows-most-47-men-famous-declaration-inde.

Knife Rights (2016, 13 June). Federal Ivory Ban Rule Goes Into Effect July 6, 2016. Kniferights.org. Retrieved from kniferights.org/legislative-update/federal-ivory-ban-rule-goes-into-effect-july-6-2016.

Kramer, R., Sawyer, R., Amato, S., and LaFontaine, P. (2017, July). The US Elephant Ivory Market: A New Baseline. *TRAFFIC Report*. Retrieved from traffic.org/site/assets/files/1378/traffic_us_ivory_report_2017.pdf.

Kukreti, I. (2019, 13 May). Israel Moots Proposal to Regulate Trade in Mammoth Ivory. *Down To Earth*. Retrieved from downtoearth.org.in/news/wildlife-biodiversity/israel-moots-proposal-to-regulate-trade-in-mammoth-ivory-64493.

Kumar, S. (2018, 6 April). A One-Man Fossil Rescue Mission. *Science, 360*(6384), 24.

Larson, Neal L., Stein, Walter, Triebold, Michael, and Winters, George (2014). What Commercial Fossil Dealers Contribute to the Science of Paleontology. *The Journal of Paleontological Sciences, 7*. Retrieved from aaps-journal.org/pdf/Contibutions-to-Paleontology.pdf.

Larson, P. and Donnan, K. (2014). *Rex Appeal: The Amazing Story of Sue, the Dinosaur That Changed Science, the Law, and My Life*. Invisible Cities Press LLC.

Larson, P. and Russell, D. (2014, April). The Benefits of Commercial Fossil Sales to 21st-Century Paleontology. Article number 17.1.2E. *Palaeontologia Electronica, 17*(1). Retrieved from palaeo-electronica.org/content/2014/739-commentary-benefits-of-fossil-sales.

Lazerwitz, D. J. (1994, Spring). Bones of Contention: The Regulation of Paleontological Resources on The Federal Public Lands. *Indiana Law Journal, 69*(2),601-636.

Lempriere, M. (2017, 28 March). Baltic Gold Rush: Illegal Amber Mining in Russia and Ukraine Is Destroying the Environment. *The Verdict*. Retrieved from verdict.co.uk/baltic-gold-rush-illegal-ambermining-russia-ukraine-destroying-environment.

Lundgren, G. (1998) Protecting Federal Fossils from Extinction. *Boston College Environmental Affairs Law Review, 26*(1), 225-262.

Lynch, J. (2018, 8 March). Russell Crowe Is Auctioning Off a $35,000 Dinosaur Skull That He Bought from Leonardo DiCaprio. *Business Insider*. Retrieved from businessinsider.com/russell-crowe-auctions-off-35000-dinosaur-skull-leonardo-dicaprio-2018-3.

Malmsheimer, R. W. and Hilfinger, A. S. H. (2003, Spring), In Search of a Paleontological Resources Policy for Federal Lands. *Natural Resources Journal, 43*(2).

Manjarrez, C., Rosenstein, C. Colgan, C., and Pastore, E. (2008). *Exhibiting Public Value: Museum Public Finance in the United States (IMLS-2008-RES-02)*. Washington, DC: Institute of Museum and Library Services.

Martill, D. (2011, 10 November). Protect—and Die. *Geoscientist 21*. The Geological Society. Retrieved from geolsoc.org.uk/Geoscientist/Archive/November-2011/Protect-and-die.

Martin, T. (2019, 12 February). Protecting our Paleontological Heritage on U.S. Federal Lands. *Paleontological Society*. Retrieved from paleosoc.org/protecting-our-paleontological-heritage-on-u-s-federal-lands.

May, P. (2017, 30 March). Look What Else They've Been Stealing from Our National Parks. *The Mercury News*. Retrieved from mercurynews.com/2017/03/30/look-what-else-theyve-been-stealing-from-our-national-parks.

Mayor, A. (2007). Fossils in Native American Lands Whose Bones, Whose Story? Fossil Appropriations Past and Present. Paper presented at the History of Science Society annual meeting, 1-2 November 2007, Washington, D.C. Retrieved from web.stanford.edu/dept/HPS/Mayorwhosebones.pdf.

Meigs, D. (2016, 22 December). Ice Age Tusks vs. Blood Ivory. *Omaha Magazine*. Retrieved from omahamagazine.com/2016/12/22/302833/ice-age-tusks-vs-blood-ivory.

Morocco Probes Dinosaur Tail Sold in Mexico Auction (2018, 21 January). Al Arabiya. Retrieved from english.alarabiya.net/en/variety/2018/01/21/Morocco-probes-dinosaur-tail-sold-in-Mexico-auction.html.

Murray [Mary Ann] and Murray [Lige M.] v. BEJ Minerals, LLC, and RTWF LLC (2020, 20 May). OP 19-0304, 2020 MT 131. Supreme Court of the State of Montana.

Murray [Mary Ann] and Murray [Lige M.] v. BEJ Minerals, LLC, and RTWF LLC (2018, 6 November May). No. 16-35506, 908 F.3d 437. United States Court of Appeals, 9th Circuit.

New Species of Ichthyosaur Discovered in Museum Collection (2020, 3 July). *GeologyPage.com*. Retrieved from geologypage.com/2020/07/new-species-of-ichthyosaur-discovered-in-museum-collection.

Nieoczym, A. (2016, 19 September). Belgian Tourist Fined $4K for Stealing Burgess Shale Fossil from Yoho National Park. *CBC Radio-Canada/CBC News*. Retrieved from cbc.ca/news/canada/british-columbia/tourist-fossil-fined-burgess-shale-stealing-1.3766674#.

Nkala, O. (2016). Chinese Demand Drives Zambia's 'Amateur' Ivory Gangs. Oxpeckers.org. Retrieved from oxpeckers.org/2016/04/chinese-demand-drives-zambias-amateur-poaching-gangs.

Nuwer, R. (2019, 5 September). Mammoth Ivory May Be Bad for Elephants—But Should It Be Banned? *Popular Science*. Retrieved from popsci.com/should-mammoths-be-protected-alongside-elephants.

Ollivier, T. (2018, 29 August). Au Maroc, le Discret Marché du Trafic de Fossiles. *Libération*. Retrieved from liberation.fr/planete/2018/08/29/au-maroc-le-discret-marche-du-trafic-de-fossiles_1675219.

Orlean, S. (1998). *The Orchid Thief: A True Story of Beauty and Obsession*. New York: Random House.

Osborne, L. The Fossil Frenzy. *New York Times Magazine*. Retrieved from archive.nytimes.com/nytimes.com/library/magazine/home/20001029mag-fossil.html.

Ostler, J. (2020, 8 February). The Shameful Final Grievance of the Declaration of Independence. *The Atlantic*. Retrieved from theatlantic.com/ideas/archive/2020/02/americas-twofold-original-sin/606163.

Pagan, D. (2008, 19 June). Stolen Fossils Returned Thanks to U of C Prof. The Gauntlet. Retrieved from Retrieved from web.archive.org/web/20140306012631/thegauntlet.ca/story/stolen-fossils-returned-thanks-u-c-prof.

Pantuso, Phillip (2019, 17 July). Perhaps the Best Dinosaur Fossil Ever Discovered. So Why Has Hardly Anyone Seen It? *The Guardian*. Retrieved from theguardian.com/science/2019/jul/17/montana-fossilized-dueling-dinosaurs-skeletons-dino-cowboy.

Parry, W. (2012, 28 June). More Dinosaur-Smuggling Cases May Follow on Tyrannosaur's Heels. *Live Science*. Retrieved from livescience.com/21263-mongolian-fossils-easily-bought.html.

Pickrell, J. (2018, 1 June). Carnivorous-Dinosaur Auction Reflects Rise in Private Fossil Sales. *Nature*. Retrieved from nature.com/articles/d41586-018-05299-3.

Pickrell, J. (2020, 15 February). The Curious Case of the Stolen Dino-Swan. *New Scientist*, 40-43.

Pringle, H. (2013, 19 November). Not Sold! 'Dueling Dinos' Flop at Auction. *Science.* Retrieved from sciencemag.org/news/2013/11/not-sold-dueling-dinos-flop-auction.

Pringle, H. (2014, 24 January). Selling America's Fossil Record. *Science, 343*(6169), 364-367.

Proctor, J. (2016, 29 February). Smuggling dinosaur fossils into U.S. costs B.C. man $25,000 US. CBC/Radio-Canada. Retrieved from cbc.ca/news/canada/british-columbia/smuggling-dinosaur-fossils-into-u-s-costs-b-c-man-25-000-us-1.3469456#.

Rayfield, E. J., Theodor, J. M., and Polly, D. (2020, 21 April). Fossils from Conflict Zones and Reproducibility of Fossil-Based Scientific Data. Society for Vertebrate Paleontology. Retrieved from vertpaleo.org/GlobalPDFS/SVP-Letter-to-Editors-FINAL.aspx.

Rare Dinosaur Skeleton Sells for €2 Million at Eiffel Tower Auction. (2018, 5 June). *France24.com.* Retrieved fromfrance24.com/en/20180605-france-dinosaur-rare-skeleton-sells-2-million-eiffel-tower-aguttes-auction-paleontology.

Report on How Mammoth Ivory Contributes to Elephant Poaching (2019, February). New York City Bar, Animal Law Committee. New York, NY: The Association of the Bar of New York City. Retrieved from s3.amazonaws.com/documents.nycbar.org/files/2018479-Elephant_Poaching_Mammoth_Ivory.pdf.

Reynolds, Matt (2018, 21 June). The Dinosaur Trade: How Celebrity Collectors and Glitzy Auctions Could Be Damaging Science. *Wired.co.uk.* Retrieved from wired.co.uk/article/dinosaur-t-rex-auction-sale-private-fossil-trade.

Riehl, F. (2015, 23 January). Government Ivory Bullies Nail First Victims, Where Else? New York. *Ammoland.com.* Retrieved from ammoland.com/2015/01/government-ivory-bullies-nail-first-victims-where-else-new-york/#ixzz-6RYOiPtem.

Rinat, Z. (2018, 14 November). In Bid to Save the Elephants, Israel and Kenya Call for Supervision of Trade in Mammoth Ivory. *Haaretz.com.* Retrieved from haaretz.com/israel-news/.premium-to-save-elephants-israel-and-kenya-seek-curbs-in-mammoth-ivory-trade-1.6655531?.

Santucci, V. (2020, 30 April). The History of Paleontology in the National Park Service: 1966-2008—Filling the Gaps in the Fossil Record. National Park Service, 9. Retrieved from nps.gov/subjects/fossils/history-of-paleontology-in-the-national-park-service.htm.

Schlossberg, S., Chasel, M. J., Gobush, K. S., Wasser, S. K., and Lindsay, K. (2020). State-space models reveal a continuing elephant poaching problem in most of Africa. *Nature: Scientific Reports*, 10:10166. Retrieved from doi.org/10.1038/s41598-020-66906-w.

Senate Report 115-415 of the Committee on Commerce, Science, and Transportation (2018, 5 December). Allowing Alaska to Improve Vital Opportunities in the Rural Economy Act, 115th Congress (2017-2018). Washington, DC: U.S. Government Publishing Office. Retrieved From congress.gov/115/crpt/srpt415/CRPT-115srpt415.pdf.

Shimada, K., Currie, P. Scott, E., and Sumida, S. (2014, March). The Greatest Challenge to 21st Century Paleontology: When Commercialization of Fossils Threatens the Science. Article number 17.1.1E. *Palaeontologia Electronica, 17*(1), 1-4. Retrieved from palaeo-electronica.org/content/2014/691-great-threat-in-21st-century.

Sicree, A. (2009, March/April). Morocco's Trilobite Economy. *ARAMCO World.* Retrieved from archive.aramcoworld.com/issue/200902/morocco.s.trilobite.economy.htm.

Simons, L. M. (2005, May). Fossil Wars. *National Geographic, 207*(5), 48-69.

Society for Vertebrate Paleontology (2020, 21 April). On Burmese Amber and Fossil Repositories: SVP Members' Cooperation Requested! *Vertpaleo.org.* Retrieved from vertpaleo.org/Society-News/SVP-Paleo-News/Society-News,-Press-Releases/On-Burmese-Amber-and-Fossil-Repositories-SVP-Memb.aspx.

Sosnowskia, M.C., Knowles, T. G., Takahashi, T., and Rooney, N. J. (2019). Global Ivory Market Prices Since the 1989 CITES Ban. *Biological Conservation, 237*, 392-399.

Standing Rock Sioux Tribe (2015, 3 February). *Title XXXVIII - (38) Paleontology Resource Code.* Retrieved from standingrock.org/content/title-xxxviii-38-paleontology-resource-code.

Stokstad, E. (2005, 2 December). Best Archaeopteryx Fossil So Far Ruffles a Few Feathers. *Science, 310*(5753), 1418-1419.

Stucke, M. E. and Ezrachi, A. (2017, 15 December). The Rise, Fall, and Rebirth of the U.S. Antitrust Movement. *Harvard Business Review.* Retrieved from hbr.org/2017/12/the-rise-fall-and-rebirth-of-the-u-s-antitrust-movement.

Sutton, B. (2018, 20 December). What's behind the Roaring Market for Dinosaur Fossils. *Artsy.net.* Retrieved from artsy.net/article/artsy-editorial-roaring-market-dinosaur-fossils.

Switek, B. (2015, 16 September). Dinosaur Skeletons Aren't Décor—They Shouldn't Be Sold to the Highest Bidder. *The Guardian.* Retrieved from theguardian.com/commentisfree/2015/sep/16/wrong-auction-dinosaur-skeletons-allosaurus-fossil.

Switek, B. (2018, 17 April). Paleontology and Private Fossil Collecting Can Be at Odds in the Hills of Wyoming. *Audubon.org.* Retrieved from audubon.org/news/paleontology-and-private-fossil-collecting-can-be-odds-hills-wyoming.

Timmins, B. (2019, 8 August). What's Wrong with Buying a Dinosaur? *BBCNews.* Retrieved from bbc.com/news/business-48472588.

Triebold, Mike (2007, 7 March). Fossils: New Journal Will Oppose Illegal Trade. *Nature, 446*(136). Retrieved from doi.org/10.1038/446136c.

US Fossils Dealer Jailed for Dinosaur Smuggling (2014, 4 June). *BBC News.* Retrieved from bbc.com/news/world-us-canada-27691816.

Vincent L. Santucci, National Park Service (2020, 22 April). National Park Service Meet a Paleontologist Series. Retrieved from nps.gov/articles/meetapaleontologist-vincentlsantucci.htm.

Webster, D. (2009, April). The Dinosaur Fossil Wars. *Smithsonian Magazine.* Retrieved from smithsonianmag.com/science-nature/the-dinosaur-fossil-wars-116496039.

Weir, F. (2001, 30 July). Skulduggery among Russia's Old Bones: Disappearing Fossils Leave a Trail of Unanswered Questions. *Christian Science Monitor.* Retrieved from csmonitor.com/2001/0730/p1s3.html.

Williams, P. (2013, 28 January). Bones of Contention. *The New Yorker, 88*(45). Retrieved from newyorker.com/magazine/2013/01/28/bones-of-contention-paige-williams.

Williams, P. (2018a). *The Dinosaur Artist: Obsession, Science, and the Global Quest for Fossils.* New York: Hachette Books.

Williams, P. (2018b, 24 September). The Fossil Wars: On the Battle Between Paleontologists and Amateur Dealers. *Literary Hub.* Retrieved from lithub.com/the-fossil-wars-on-the-battle-between-paleontologists-and-amateur-dealers.

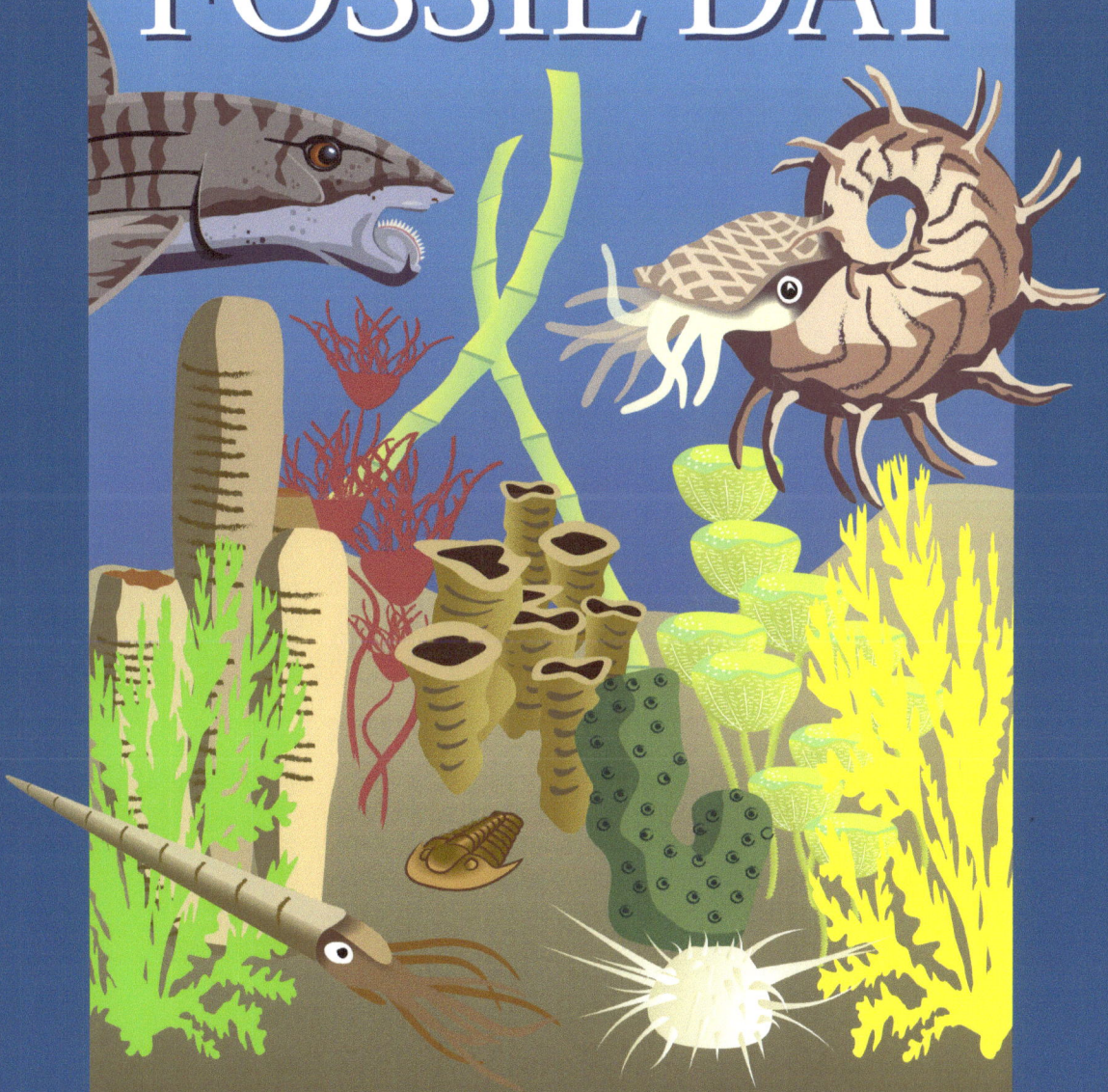

"Cannibal the Animal"—The Making of a Monster

Mats E. Eriksson

A hellish monstrosity of an animal—like a beastly entity straight out of nightmares—has come to sculptural life. And it has Death Metal, the primordial ooze, and Alex Webster, the mighty bass giant of Cannibal Corpse, Blotted Science, and Conquering Dystopia, written all over it!

In 2017, a new and gigantic fossil polychaete worm, *Websteroprion armstrongi*, was discovered and unveiled to the world (Eriksson, Parry & Rudkin, 2017). The creature is an ancestor to the extant marine Bobbit worms—ambush predators that hunt in stealth mode for octopuses and fish. The fossil species was discovered in 400-million-year-old rocks from the Devonian Period in Canada and was named in honor of Webster.

Websteroprion armstrongi possessed the largest worm jaws recorded in the entire fossil record, reaching over one centimeter in length (fossil worm-jaw elements, known as scolecodonts, usually range no larger than a few millimeters). Investigation of the relationship

The monster sculpture in progress with its "daddy," model maker Esben Horn. At this stage, the worm body had been roughly sculptured in Styrofoam and the huge jaws still needed adjustment. © Mats E. Eriksson; used by permission.

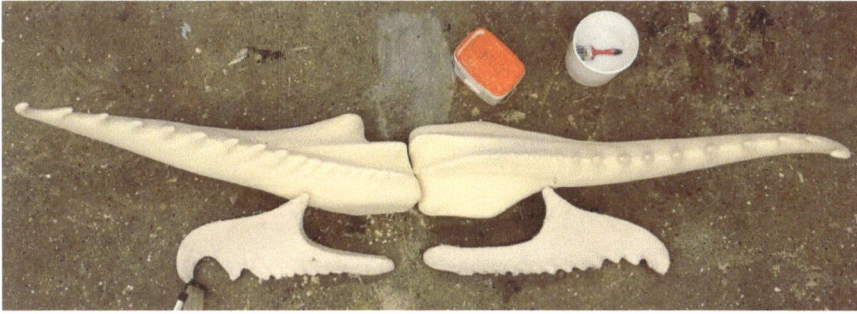

Above: Sections of the jaws of *Websteroprion armstrongi* before being mounted. *Bottom:* The completed sculpture in all its glory, sporting its slick black leather suit. © Esben Horn; used by permission. *Inset*: Rock Fossils logo on a poster from a 2015 presentation at the Berne Natural History Museum in Switzerland.

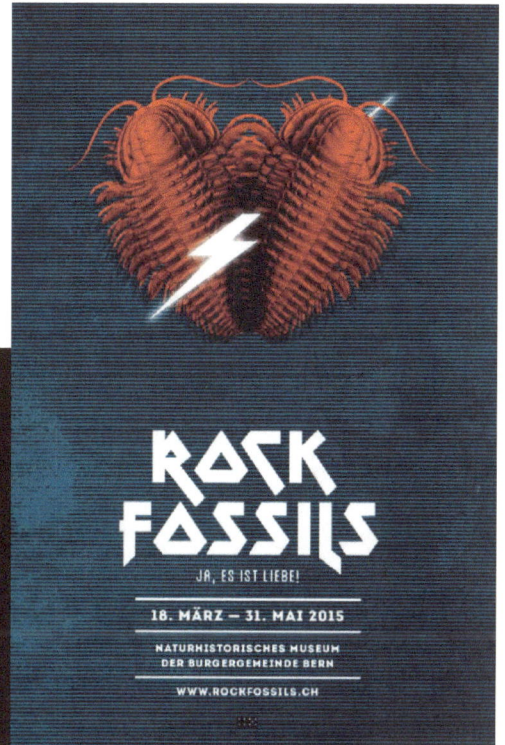

A reconstruction of the fossil Bobbit worm *Websteroprion armstrongi*, from the Devonian Kwataboahegan Formation of Ontario Canada, attacking an extinct bony fish. Art © James Ormiston, selected as one of the 2017 PLOS Paleontology Top 10 Open Access Fossil Taxa.

between body and jaw size suggested that this animal achieved a total body length in excess of a meter, which is comparable to that of modern "giant eunicid" worm species (large, exclusively marine polychaetes).

While this is certainly impressive, why settle for big when you can go mega? Hence, the sculptural reconstruction of the animal, modeled by

the skilled hands of prehistoric sculpture artist extraordinaire, Esben Horn, at his company 10 Tons, in Copenhagen, Denmark, measures over two and a half meters in its majestic height and shows the anterior (head) portion of the beast as if it was

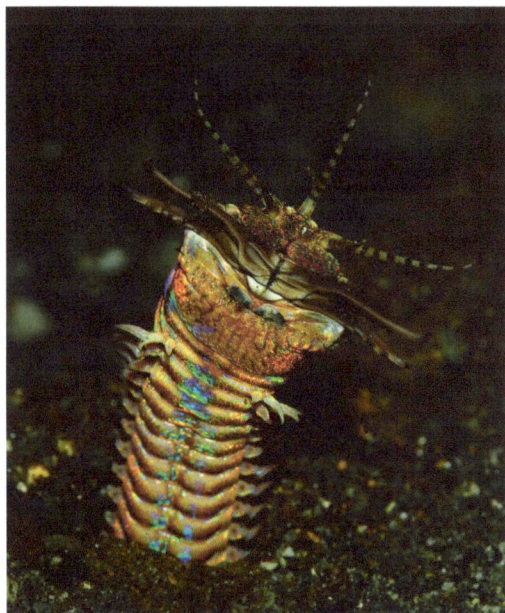

emerging from the sea floor, jaws spread wide apart and waiting to drag some passing prey below where it can be devoured. The final result brings to mind something from the cult flick *Tremors* (1990), which happens to be one of Esben Horn's favorite films.

The jaws of the giant sculpture were modeled after the actual fossils, sculpted in polyurethane foam, and then covered in a layer of fiberglass-reinforced acrylic gypsum. The outermost layer was modeled in "Rasmonite" (an acrylic gypsum substance modified by Rasmus Frederiksen of 10 Tons), which turns rock-solid upon hardening, and finally painted. The worm body of this "cannibal animal" was modeled after those of its now-living relatives (as no fossilized soft tissue has been found) and carved out of Styrofoam. A black leather suit was tightly fitted over the finished sculpture, giving it not only a unique look, but underscoring the metal reference in the species name and metal aesthetics in general. The finished beast is a sharply dressed, evil worm ready for a night out of serious ambush hunting.

The new worm monstrosity is the centerpiece of the hugely successful traveling exhibition *Rock Fossils*. This exhibition, organized by an international group of scientists and artists, Jesper Milán (Denmark), Esben Horn (Denmark), Rune Fjord (Denmark), Achim G. Reisdorf (Germany), and me (Sweden), portrays fossils named after rock stars and tells their stories: the discoveries, the musicians, the scientists and, of course, the primordial organisms themselves. The *Websteroprion* sculpture was shown to the public for the first time when the Rock Fossils exhibition opened in June 2018 in Luxembourg.

Work Cited

Eriksson, M.E., Parry, L., and Rudkin, D.M. (2017). Earth's Oldest 'Bobbit Worm'—Gigantism in a Devonian Eunicidan Polychaete. *Scientific Reports, 7,* 43061. Retrieved from doi:10.1038/srep43061.

Above: The *Websteroprion armstrongi* sculpture was a collaborative effort between sculptor Esben Horn (left) who took care of all artistic aspects, and Mats Eriksson, who mainly acted as a scientific advisor and design playmate. © Magnus Eriksson; used by permission. *Left:* A modern polychaete, the marine worm, *Eunice pennata.* © Arne Nygren, licensed by Creative Commons.

Earth's Oldest Bobbit Worm

As far as Bobbit worms go, *Websteroprion armstrongi* was pretty massive, and for a polychaete worm in the Devonian, even more impressive. Wide open, its jaws would have spanned about two cm (not quite an inch) across. The body, though not preserved, is estimated by the authors to have been more than a meter (three feet) in length.

Dr. Mats Eriksson, who described the new species, explained, "Gigantism in animals is an alluring and ecologically important trait, usually associated with advantages and competitive dominance. It is, however, a poorly understood phenomenon among marine worms and has never before been demonstrated in deep time based on fossil material in this group of animals. The new species demonstrates a unique case of polychaete gigantism in the Paleozoic, some 400 million years ago." Eriksson adds, "The specific driving mechanism for *W. armstrongi* to reach such a size remains ambiguous."

Would *Websteroprion* have been an ambush predator like its modern-day relatives? "We have no empirical evidence of its diet," Dr. Eriksson explains. "As we lack soft parts, we don't have access to preserved gut contents. Inferring the diet of extinct worms (even jaw-bearing ones) is difficult. Especially considering that there are jaw-bearing extant forms that, despite looking like 'fierce' carnivorous predators, have proven to have a wide range of feeding habits. With that being said, given its size and compared to its closest modern relatives, I would assume that *W. armstrongi* had a similar mode of life and feeding habit as the modern Bobbit worms. So, Devonian fish and cephalopods may not have been safe from this creature."

Websteroprion specimens caught the eyes of the authors in the collection of the Royal Ontario Museum in Toronto, Canada, but they had actually been collected more than twenty years earlier. "The fossil specimens were collected over the course of a few hours in a single day in June 1994, when Derek K. Armstrong of the Ontario Geological Survey was dropped by helicopter to investigate the rocks and fossils at a remote and temporary exposure in Ontario," Eriksson explains. "Sample materials, from what proved to belong to the Devonian Kwataboahegan Formation, were brought back to the Royal Ontario Museum in Toronto, Canada, where they were stored until they came to our attention.

"Luke Parry [then a PhD student at the University of Bristol] was doing guest research on full-body polychaete fossils at the Royal Ontario Museum at the time, and Dave Rudkin showed him the specimens. Luke took a quick photograph and sent it to me, knowing that I am an expert on this fossil group."

Dr. Eriksson continues, "I was quite disappointed when I first laid eyes on the photographs. The preservation was far from exceptional, and at first I concluded that it wasn't worthwhile pursuing. But then I asked about the size! The original image didn't come with a scale bar, and I had simply assumed that the specimen was the 'standard' millimeter size. I asked Luke, who said that they were, in fact, pretty big, and provided the scale.

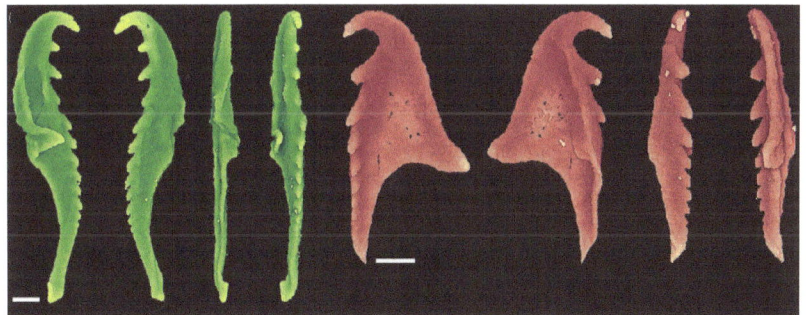

Above: CT-scans of tiny jaw elements (scolecodonts) of *Websteroprion armstrongi*, the new Devonian marine worm described by Eriksson, Parry, and Rudkin in 2017.

That's when I strongly suspected that these must be the largest fossil polychaete jaws ever reported, a hunch that subsequent research confirmed. That certainly whet my wormy appetite!"

He adds, "Our study is an excellent example of the importance of looking in remote and unexplored areas for finding new exciting things, but also the importance of scrutinizing museum collections for overlooked gems."

Excerpted from "Top 10 Open Access Fossil Taxa of 2017: *Websteroprion armstrongi*" by Ian Hamilton. *The Official PLOS Blog,* 5 December 2017.

Book Review

by Spencer G. Lucas

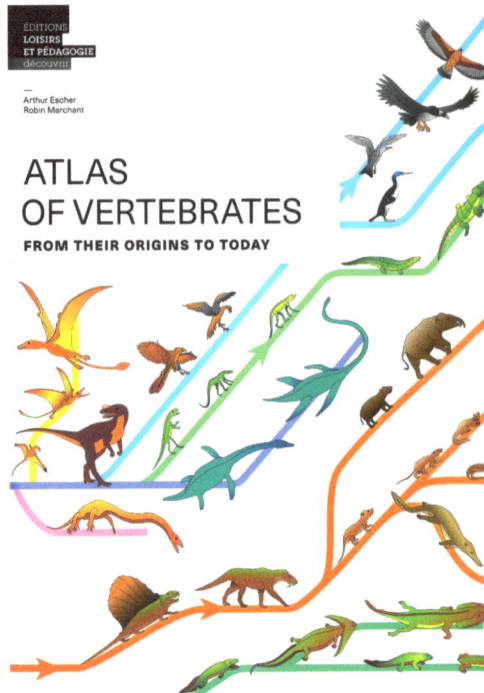

Spread: The evolutionary tree of the sauropods. *Facing page, top*: a detail of the fold-out poster that combines the family trees of all the vertebrates.

Atlas of Vertebrates from Their Origins to Today by Arthur Escher and Robin Marchant. Le Mont-sur-Lausanne, Switzerland: Éditions Loisirs et Pédagogie, 2019, 40 pages. Available in French and English editions.

During the last half century, the bar for prehistoric art has been set high. This is particularly true of dinosaur art, now being created by a vast number of talented artists from all over the globe. For a prehistoric art book to stand out from "the pack," then, is unusual. This book stands out. It does so because it is unique among prehistoric art books as a collection of evolutionary trees (phylogenetic trees or phylogenies), densely illustrated by gifted artist Arthur Escher, who, incidentally, is the son of M. C. Escher (1898-1971), famed for his mathematically inspired graphic artwork. Furthermore, as a teaching tool, this book has few equals.

The book's front endpapers begin with a "generalized tree of life" and a key to understanding the evolutionary trees that follow. Those trees run horizontally across this large-format volume (12.25" wide and 12.5" tall). There are a total of fifteen evolutionary trees: (1) fish and their emergence from water; (2) sharks, rays and chimeras; (3) diversification of bony fish; (4) the first reptiles; (5) the flight of reptiles; (6) terrestrial dinosaurs; (7) the return of the reptiles to water; (8) the flight of feathered dinosaurs; (9)

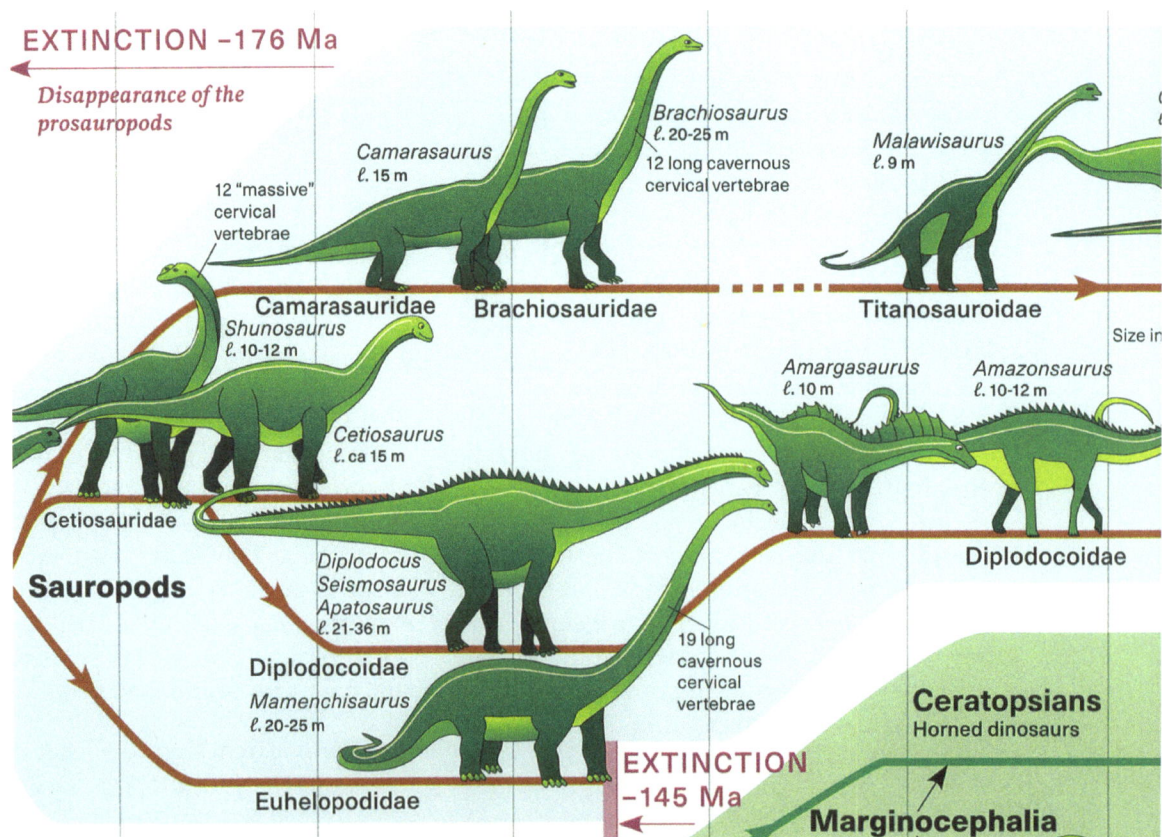

accompany each evolutionary tree. Some of the trees end in numbers that guide the reader to the base of the succeeding evolutionary tree elsewhere in the book. For example, within the larger tree of early mammals, the tree of afrotherian mammals ends at number 11 (afrotheres belong to groups that are either currently living in Africa or are of African origin: golden moles, elephant shrews, tenrecs, aardvarks, hyraxes, elephants, sea cows, and several extinct clades). All the trees come together in a poster in a sleeve at the back of the book which, when unfolded, measures about two feet by three feet.

the diversification of reptiles; (10) early mammals; (11) elephants and sea cows; (12) primates; (13) the tribe of bipedal primates; (14) carnivores; and (15) ungulates and cetaceans. The relevant portion of the geological time scale is on the horizontal axis beneath each evolutionary tree. The last two-page layout illustrates the major phases of extinction from the end of the Ediacaran (~ 541 million years ago) to the ongoing extinction of the Modern World.

Each evolutionary tree is illustrated with dozens of color drawings of key animals along the branches, identified by genus name with a metric estimate of total body length. All generic names are indexed at the back of the book. A brief explanatory text and instructive diagrams

The trees in the book, of course, are not all above scientific discussion—there is ongoing debate about many aspects of the evolutionary history of the vertebrates. But these are up-to-date trees that mostly capture a consensus in good, current scientific thinking on the phylogeny of a vertebrate group.

Beyond its unique content, this well-illustrated, easy-to-understand volume is a remarkable teaching tool. Darwin's *On the Origin of Species* contained only one illustration, a generalized evolutionary tree. Such trees have long been the graphic by which paleontologists have depicted evolutionary history—the unfolding diversification of life over geologic time—and show, to borrow again from Darwin, "descent with modification," i.e., the evolutionary process.

Atlas of Vertebrates is for anyone with a keen interest in vertebrate evolution, and there's no better gift for younger people, say middle-school aged or older, who want to grasp the main points of this long and fascinating process.

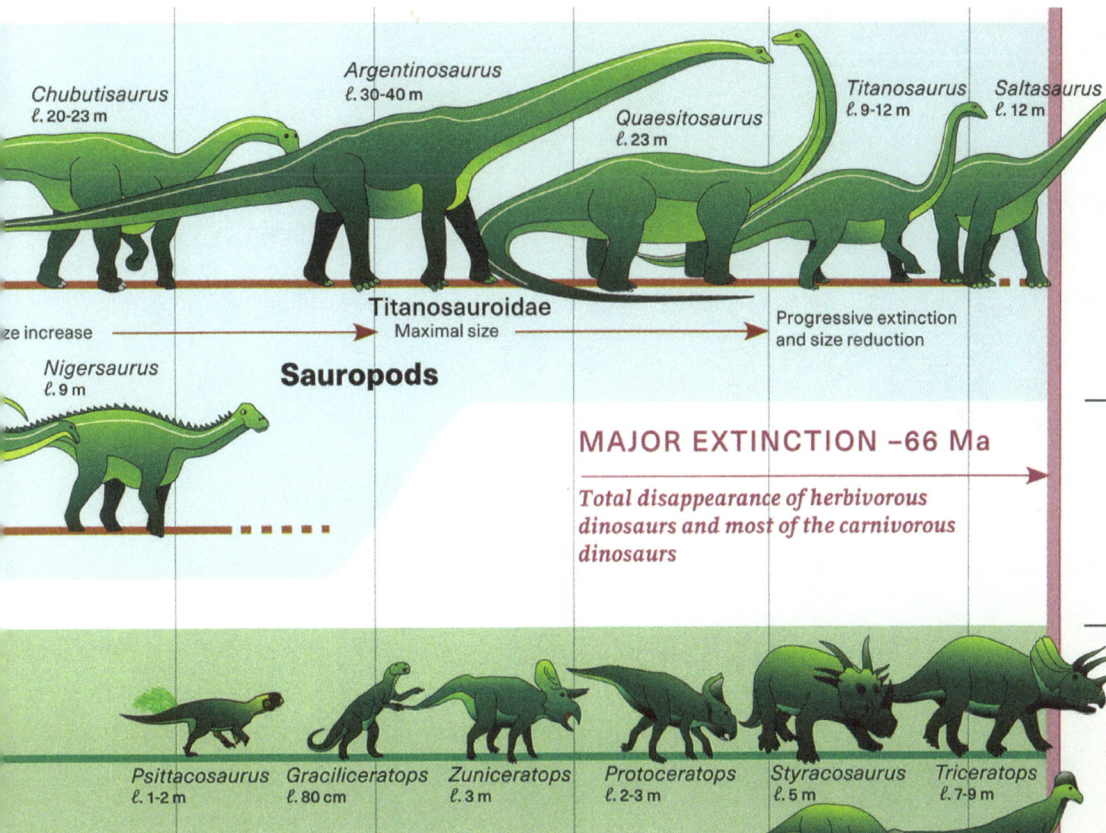

Chubutisaurus ℓ. 20-23 m

Argentinosaurus ℓ. 30-40 m

Quaesitosaurus ℓ. 23 m

Titanosaurus ℓ. 9-12 m

Saltasaurus ℓ. 12 m

size increase

Nigersaurus ℓ. 9 m

Titanosauroidae Maximal size

Progressive extinction and size reduction

Sauropods

MAJOR EXTINCTION −66 Ma

Total disappearance of herbivorous dinosaurs and most of the carnivorous dinosaurs

Psittacosaurus ℓ. 1-2 m

Graciliceratops ℓ. 80 cm

Zuniceratops ℓ. 3 m

Protoceratops ℓ. 2-3 m

Styracosaurus ℓ. 5 m

Triceratops ℓ. 7-9 m

Political Pangaea: Ancient Continent, Modern Borders
A map by Massimo Pietrobon

Political Pangaea

Massimo Pietrobon

I love geography and I spend a lot of time playing with maps and graphic images. I started to create this map of Political Pangaea just for fun, but I soon began to realize the potential of the idea: the union of modern nations into a single continent was a return to an original condition in which conventional separations disappeared.

Geographical maps have enormous symbolic power. What is in the center is important; the rest is a second thought. What is on top is more important than what is on the bottom. The deformations of proportional sizes are also conceptually powerful. When these implications are graphically represented in a world map, they crystallize in people's minds and that version of the map becomes the way people commonly think about the world.

In creating Political Pangaea, I tried to be as accurate as possible. I did a great deal of research into maps of tectonic plates and movement as well as theories regarding Pangaea and how the land masses appeared in those times. I superimposed layers of maps that illustrated existing theories, and then I drew my map of Pangaea carefully in Photoshop.

Beyond the formal process of superimposing national borders onto the ancient continent of Pangaea, the "Political Pangaea" project involved theoretical (and metaphysical) considerations.

Reuniting the Earth in a single expanse of land represents a return to the unity of the planet and of humankind, a defiant rejection of the divisions that have proven so convenient to the people who govern us.

The result is fun because it contains some geopolitical shocks. The United States now finds itself looking the Arab states in the face, with Colombia just downstairs and Cuba nearly touching Miami!

We Europeans, in contrast, finally have Africa at our front door. No need any longer for thousands of human lives to be lost at sea in order to reach our shores—they can come by bicycle!

Black Americans are finally reunited with their African cousins, and if they want to visit each other, they can take the bus.

Not only that, Moroccans can finally walk to Quebec! It was about time.

I realized that the first and more important meaning of Political Pangaea is that the entire world is united. Human beings are all equal dwellers of the same planet.

One world.

One humanity.

Long live Political Pangaea!

A Pangaea Primer: Pangaea was the earliest of supercontinents in Earth's history, though some believe others existed even earlier. Pangaea began developing over 300 million years ago and was surrounded by the enormous ocean known as Panthalassa. Similar to parts of Central Asia today, the center of the landmass is thought to have been arid and inhospitable, with temperatures reaching 113° F (45° C). The extreme temperatures revealed by climate simulations are supported by the fact that very few fossils are found in the modern-day regions that once existed in the middle of Pangaea. The strong contrast between the Pangaea supercontinent and Panthalassa is believed to have triggered intense cross-equatorial monsoons. Plants and animals nevertheless spread across the landmass, and animals (such as dinosaurs) were able to wander freely across the entire expanse of Pangaea. Around 200 million years ago, magma began to swell up through a weakness in the earth's crust, creating the volcanic rift zone that would eventually cleave the supercontinent into pieces. Over time, this rift zone would become the Atlantic Ocean. (Based on original material by Nick Routley; used by kind permission of VisualCapitalist.com.)

ADVENTURES IN PALEONTOLOGY
FROM CHICAGO

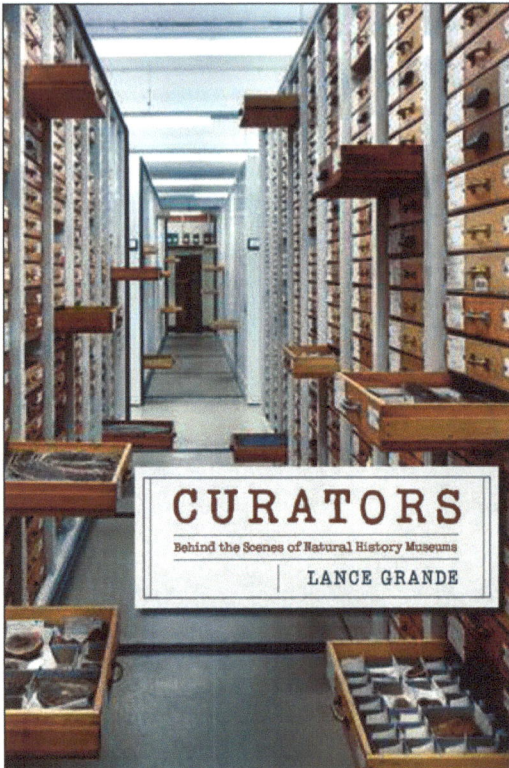

CURATORS

Behind the Scenes of Natural History Museums

Lance Grande

"I was left with a feeling of slightly abashed awe after reading about the stars of *Curators*, Grande's book on the life scientific at Chicago's Field Museum of Natural History. . . . Far from the popular image of introverted specialists tending drawers deep in the vaults, these curators are Indiana Jones figures swashing and buckling their way to remote regions, dealing with drug barons or cantankerous farmers as needs must, bent on returning with scientific treasure."—Richard Fortey, *Nature*

CLOTH $35.00

THE LOST WORLD OF FOSSIL LAKE

Snapshots from Deep Time

Lance Grande

"Enchanting. . . . *The Lost World of Fossil Lake* is a splendidly illustrated compendium on the c.52 million-year-old fossils found at the lakebed, written with grace and authority. . . . With its 243 fine colour plates, it will appeal to both amateur and professional geologists and palaeontologists."—*Times Higher Education*

CLOTH $45.00

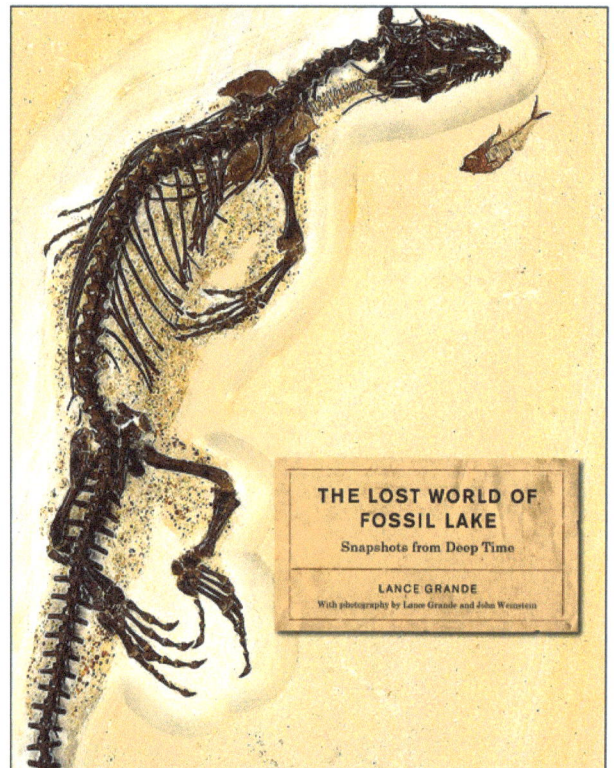

The University of Chicago Press www.press.uchicago.edu

The Sam Noble Oklahoma Museum of Natural History and Evidence for an Ordovician Mass Extinction Event

Albert J. Copley

The Sam Noble Oklahoma Museum of Natural History, located on the campus of the University of Oklahoma in Norman, Oklahoma, is not only one of the nation's nicer museums, it houses significant paleontology collections. Established by the Territorial Legislature of Oklahoma seven or eight years before statehood came in 1907, the Sam Noble is actually older than the state of Oklahoma itself. For a long time the museum languished without much development and then, in 1999, almost exactly 100 years after the original founding, the museum opened in a beautiful new building.

The museum's exhibits include many non-paleontology-related offerings, as well as classes and meetings on a variety of subjects and for all ages of people, from the very young through adults. For obvious reasons, the museum is closed as of this writing, though a virtual visit, classroom activities, educational speakers, and even coloring pages and movies are available on their site: https://samnoblemuseum.ou.edu. The Sam Noble hopes to reopen to the public late in the summer of 2020, but it would be a good idea to verify before you visit: call 405.325.4712 or check announcements on the museum's homepage.

Our focus here, though, is on the fossils. The Sam Noble Museum's invertebrate fossil collection amounts to one million or more specimens. Donations have been received from the School of Geology and Geophysics of the University of Oklahoma, from Sam Noble Museum staff members, from the Oklahoma Geological Survey, and from the British Petroleum Company and Amoco. Major collections highlight all fifty states of the USA as well as fifty or more other countries. Paleozoic Era collections are a special part of the museum's holdings, and fossils from Canada, Britain, Sweden, the Czech Republic, Russia, and Poland are guaranteed to astonish. Cambrian and Ordovician trilobites are particularly well represented, and insights into their biostratigraphy, paleoecology, and macro-evolutionary patterns are provided by specimens from Oklahoma, Texas, Missouri, the Upper Mississippi Valley, the Great Basin, and Wyoming.

The vertebrate fossil collection has made the Sam Noble Museum a major research facility in the southern plains. More than 70,000 specimens are cataloged, including Early Permian tetrapods, Jurassic dinosaurs, Miocene-Pliocene mammals of Oklahoma, and other Cretaceous vertebrates of the western interior states. Noteworthy items are the Early Permian fauna, specimens from the Upper Jurassic Morrison Formation, and species of Miocene horses. In addition, the collections of Cretaceous microvertebrates are outstanding.

To discuss all of the Sam Noble Museum's collected fossils would be impossible in a brief article, but several specific holdings deserve a special mention. One of these is a magnificent slab displaying multiple specimens of *Homotelus bromidensis*. The Middle Ordovician Bromide Formation of Carter County, Oklahoma, is world famous for its diverse echinoderm and trilobite fauna, and examples of *Homotelus bromidensis* from the Bromide may be seen in collections worldwide. After gazing at the slab for a time, I realized I could be seeing an example of *thanatocoenosis*—that is, a death assemblage or kill zone. I felt as though the fossils were trying to tell me a story.

Years ago, one of my professors in graduate school at the University of Oklahoma—Dr. R. W. Harris—would tell us: "All of the information about the earth which we would ever have was already here within these rocks or fossils. It was simply up to us to ask the right question and then listen to what the rock or fossil had to say."

My thoughts were running through the various questions. Why should many individuals, which appeared to be of the same degree of development, the same size, and the same physical condition, and were oriented with the same side up, have been preserved in this place at a particular time in the history of the world? Could sheer coincidence be an explanation? I have seen examples of these fossils in several different museums. After seeing the display in Oklahoma, I finally realized their significance and decided a few inquiries were warranted.

I immediately came to understand that the death of these fos-

Facing page: Mortality plate of *Homotelus bromidensis* trilobites, Cleveland Museum of Natural History. © Tim Evanson; used by permission.

Oklahoma fossil invertebrates

sils had been studied minutely. First, the species name "*bromidensis*" has been assigned to several different genera. At one time or another, it has been placed under *Homotelus, Isotelus, Vogdesia,* and *Anataphrus.* (For this article, I will consider it to be *Homotelus.*) Second, the Bromide Formation includes at least five different horizons from which *Homotelus bromidensis* specimens have been extracted.

Densities of *Homotelus* in the Bromide Formation have been calculated as high as 170 per square meter. The fossils seem to be entire animals and not simply molts. Mating clusters have been mentioned as a possibility for these mass occurrences, on the analogy of the mating behavior of modern-day horseshoe crabs along the Atlantic Coast, though the suggestion seems to have been discounted by most researchers. The individuals don't appear to have a preferred orientation—that is, they were not shifted by water currents such that they pointed in more-or-less the same direction. There does seem to strong evidence to call this a "trilobite kill."

Most readers will know that the Ordovician Period immediately followed the Cambrian Period, which is the oldest period of the Paleozoic Era. The Ordovician is considered to have started 485.4 +/- 1.9 million years BCE (before the Common Era). About 449.8 +/- 1.5 million years BCE, the Ordovician ended in the Taconic Orogeny, a mountain-building period that involved most of modern-day New England.

But the end of the Ordovician Period and the beginning of the Silurian Period were also marked by a major mass extinction. Seas during the Ordovician covered large percentages of the continents of the world, and Ordovician deposits can be found today high in the Himalaya Mountains. (The Himalayas were formed as the result of a continental collision during the Upper Cretaceous, and they did not exist as mountains during the Ordovician. The Paleozoic fossils found there are the result of the massive upthrusting of the ocean floor as the mountains were formed).

Rapid climate change, the cooling of the relatively warm Ordovician oceans, and the loss of habitat along the continental shelves as a result of a major sea-level drop led to the extinction of about 85% of the world's marine species at the end of the Ordovician, especially brachiopods, bivalves, echinoderms, bryozoans, and corals.

Could the masses of *Homotelus bromidensis* specimens in the Bromide Formation be evidence of an

Facing page: Homotelus bromidensis Esker 1964 (Bromide Formation, Middle Ordovician; Criner Hills, Carter County, Oklahoma). On display at the Nebraska State Museum of Natural History, Lincoln, Nebraska. Note the exposed hypostome in the central individual and in several others. © James St. John, CC License 2.0 Generic. *Above*: Some Silurian brachiopods from the Sam Noble Museum collection. From left: The spiriferid, *Delthyris kozlouiskii*; the chileid, *Dictyonella gibbosa*; and the orthid, *Ptychopleurella rugiplicata*. © Sam Noble Museum. *Below*: A Columbian mammoth from Jackson County, Oklahoma, on display at the Sam Noble Museum. © Albert Copley.

At right: Oklahoma fossil vertebrates. Top: *Eryops*, a Permian amphibian from Noble County. Despite its appearance, *Eryops* is more closely related to frogs and salamanders than to crocodilians. *Middle:* Another Permian amphibian, *Diplocaulus*, from Noble County, OK. Both © Albert Copley. *Bottom:* The skull of *Pentaceratops*, a Cretaceous dinosaur on display at the Sam Noble Museum. Findings from a 2011 study indicate that the specimen may actually be a *Titanoceratops*, a new type of dinosaur. © Allison Meier, Licensed by Creative Commons. *Facing page:* Fossilized fronds of the Pennsylvanian tree fern, *Pecopteris*, from Osage County, OK.

extinction event? Such a conclusion would require evidence of mass kills in other places contemporary with the Bromide Formation. As mentioned above, a mass extinction did take place late in the Ordovician, but extinction events didn't happen all at once and, in any case, a mass extinction is a complex event and requires a complex explanation. Perhaps the *Homotelus* "kill plates" suggest that problems within the oceans may have been occurring earlier during the Ordovician.

Exploring such a possibility is beyond the scope of this article, but the example of *Homotelus* clusters in the Bromide Formation does provide an opportunity to consider the wealth of information that may be gained from a single specimen. All we need to do is take a closer look and listen to what the fossil may be trying to tell.

References

Esker, G.C. (1964). New Species of Trilobites from the Bromide Formation (Pooleville Member) of Oklahoma. *Oklahoma Geology Notes 24*: 195-209.

Gradstein, F., Ogg, J., and Smith, A. (Eds.) (2005). *A Geologic Time Scale 2004.* Cambridge, England: Cambridge University Press.

Ross, R.J. (1970). Ordovician Brachiopods, Trilobites, and Stratigraphy in Eastern and Central Nevada. *United States Geological Survey Professional Paper 639.* Washington, CD: US Government Printing Office. Available at https://pubs.usgs.gov/pp/0639/report.pdf.

Facing page, *top:* A leaf from the Permian cycad *Taeniopteris* sp., Tillman County, OK. *Bottom:* The entrance to the Sam Noble Museum of Natural History during a special exhibit dedicated to arachnids. The Museum is located at 2401 Chautauqua Avenue, Norman, Oklahoma; (405) 325-4712. *This page*, *left:* A Pennsylvanian *Lepidodendron* from Carter County, OK, in the Sam Noble Museum's collection. © Albert Copley. *Below:* Another *Homotelus bromidensis* mortality plate from the M. Ordovician Bromide Formation, displayed at the Nebraska State Museum of Natural History. © James St. John, CC License 2.0 Generic.

Trilobite Clusters — Notes from the Literature

In south-central Oklahoma, the Pooleville Member of the Bromide Formation ... contains horizons of dense clusters of a single species of trilobite, *Homotelus bromidensis*. The lower horizon is dominated by exoskeletons that have librigenae ["free cheeks" on either side of the glabella outside of the facial sutures] and hypostomes [a hard mouthpart of trilobites found on the underside of the head] in place and are interpreted as carcasses rather than molts. The upper horizon is composed largely of molts.

In both horizons, facing directions of the exoskeletons are random, and the majority of specimens are in convex-up orientations, [implying] that the clusters do not owe their origin to transportation by waves or currents. Rather, they most likely reflect aspects of the behavior of living trilobites. All specimens are [mature individuals] and encompass a narrow size range. The trilobite horizons are overlain by sparsely fossiliferous lime mudstone or thin, barren clay layers, [suggesting] that catastrophic mud blanketing, probably during major storms, preserved the horizons.

Thus, the clusters ... most likely record behavioral aggregation of individuals for synchronous molting and reproduction.... Dense clusters of trilobites have been reported in the literature for more than sixty years, but their significance went unrecognized until Speyer and Brett (1985) published their interpretation of assemblages of *Phacops rana* from the Middle Devonian of New York State [and] realized that the preservation of numerous articulated exoskeletons required rapid burial.... They also proposed that the clustering behavior preserved in the assemblages was due to synchronous mating and molting by mature individuals. This behavior is seen in many extant marine arthropods, such as horseshoe crabs, that aggregate in large groups for mating and molting. Trilobite clusters from the Middle Ordovician Bromide Formation of Oklahoma have been quarried commercially for many years, and examples are housed in collections of universities and museums throughout North America.

Description of the Homotelus Clusters

The study site ... is located in the Dunn Quarry near Ardmore and yields an approximately forty-meter section through the upper Bromide Formation. The two horizons differ dramatically in the proportion of molts and carcasses.... In both horizons, facing directions of [individuals] are apparently random with no obvious preferred orientation, and a significant majority is in the normal, convex-up orientation. Apart from a single enrolled individual, all specimens in both horizons from the Dunn Quarry are horizontally extended. However, a large slab from the same locality in the collections of the School of Geology and Geophysics, University of Oklahoma, includes 10% enrolled individuals.

The exoskeletons of both Horizons 1 and 2 lack epibionts [i.e., organisms that live on the body surface of another] such as encrusting bryozoans. This suggests that the exoskeletons could not have been exposed for long periods at the sediment surface following death or molting.

Behavioral Aggregation

The preservation of articulated specimens, both as carcasses and molts, does imply rapid burial. Following other interpretations of trilobite clusters, the *Homotelus* clusters [may have been] preserved by mud blanketing by [storm deposits] shortly after the animals had been killed or molted. It is not possible to determine whether carcasses represent animals that were buried alive, or whether they were killed prior to burial. However, the fact that enrolled specimens are rare suggests that the animals were overcome very quickly.

There [have been] several potential explanations for trilobite clusters, including feeding, response to environmental disturbance, or mating behavior, and it is possible that high and low-density clusters represent different aspects of trilobite behavior. The large numbers of specimens in the *Homotelus* clusters, together with the presence of molts, is consistent with the hypothesis of behavioral aggregation for spawning and synchronous molting [which] may have been a characteristic of trilobites in general.

Work Cited

Speyer, S. E. and Brett, C. E. (1985). Clustered Trilobite Assemblages in the Middle Devonian Hamilton Group. *Lethaia, 18*, 85-103.

Excerpted from Karim, T. and Westrop, S. R. (2002). "Taphonomy and Paleoecology of Ordovician Trilobite Clusters, Bromide Formation, South-Central Oklahoma." *PALAIOS*, 17(4) 394-402.

The temptation may be to envision trilobites as solitary creatures, captured for eternity as fossils as they proceeded stoically and individually about their daily routine. But certain aspects of the behavior of these ancestral arthropods may not be exactly as we imagine. As paleontologists have recently begun to learn, it seems that, rather than being isolated "loners," many trilobites were highly communal animals that often lived in tightly packed groups, perhaps even traversing the world's seas in long single-file, cephalon-to-pygidium lines. This was apparently a lifestyle that provided both safety in numbers and markedly increased each trilobite's procreative opportunities.

In fact, in some cases, particularly those trilobites found in sedimentary outcrops in such diverse locations as Oklahoma, the Czech Republic, Russia, Morocco, Utah, and British Columbia, these ancient creatures were nothing less than pervasive within their given ecosystem. In these locales (as well as in an ever-increasing number of sites world-wide), layers of Paleozoic rock have been found that are literally covered in trilobites.

A number of scientists believe that these mass-mortality assemblages, such as those exhibited by the Ordovician asaphid *Homotelus bromidensis*, may reflect the end result of an oceanic tidal estuary draining or evaporating, leaving its inhabitants quite literally high-and-dry. Others state that some trilobites, such as the Devonian phacopid *Eldredgeops milleri*, may have followed a life cycle that would have drawn their species together in prolific numbers at certain times of the year to create mating assemblages.... Another scientific thought postulates that large numbers of the same trilobite species may have continually lived in close proximity, sharing a particularly hospitable ecological niche for protection and best utilization of resources.

Excerpted from "Trilobite Multiplicity" in the American Museum of Natural History's *Trilobite Files.*

At right: a pair of mass-mortality trilobite plates. Top: *Kootenia youngorum.* This specimen is the only known example of a display of communal behavior among this rare species from the Cambrian of Utah. *Bottom: Conocoryphe sulzuri.* These blind Cambrian trilobites found in the Czech Republic may have gathered together for protection. © American Museum of Natural History.

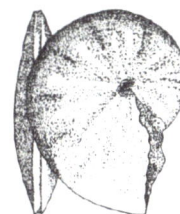

The trilobite clusters described here from the upper Tremadocian (earliest Ordovician) Fezouata Shale in Morocco are overwhelmingly dominated by *Ampyx priscus* with rare occurrences of [associated] asaphids and calymenids. [The] trilobites are preserved as internal or external molds, and ... show no remains of appendages and internal organs. In each cluster, trilobites are arranged in a linear fashion with their anterior end facing one direction and lie on the surface of a single bedding plane with the dorsal surface of their exoskeleton directed upwards. The number of *A. priscus* specimens

Accumulation layers or shell beds are frequent in the fossil record. They typically consist of disarticulated organisms and heterogeneous exoskeletal fragments assembled together and often oriented by currents. The *Ampyx* clusters from the Fezouata Shale have none of these diagnostic features. In contrast, they are made up of articulated monospecific individuals and are not associated with sedimentary structures indicative of sea bottom troughs or burrows. Moreover, the consistent anterior polarity of individuals could hardly be explained by the action of currents. [T]hese linear clusters most likely represent ...

Above: The Early Ordovician blind asaphid trilobite, *Ampyx priscus*, in single file in the Moroccan Fezouata Shale. *Facing page:* Closeup of another *A. priscus* specimen held in the Musée des Confluences in Lyon, France, showing the rare occurrence of a non-*Ampyx* trilobite in line. The last individual is the phacopid *Parabathycheilus*. Images © Vannier et al. (2019) and Scientific Reports.

in clusters varies from three to twenty-two. No specimen is disarticulated, suggesting that they represent carcasses and not exuviae. *Ampyx* specimens [are distributed] within a relatively narrow size range probably represent adult or subadult sexually mature animals. The distance between individuals is relatively short and rarely exceeds twice the body length, giving the trilobite clusters a cohesive appearance. Succeeding specimens are frequently in contact with each other via their long glabellar and genal spines. Overlapping individuals are frequent.

most of the original position of individuals at the time of their death. This interpretation is strongly supported by geological evidence. Detailed sedimentological analyses show that the Fezouata Shale is characterized by background sedimentation repeatedly disturbed by storm sequences made of normally graded, very fine sands and coarse siltstones that locally exhibit small oscillation structures, indicating that the depositional setting was located close to the storm wave base. The amount of sediment deposited during a storm event was probably sufficient to entomb trilobites and other epibenthic animals but not powerful enough to take them away. [W]ater poisoning may have also participated in rapidly killing *Ampyx* [given that] neither sedimentary disturbances nor body attitudes indicate [an attempt] of trilobites to escape burial. Many trilobites, including *Ampyx priscus* had the capacity to enroll [and] the extreme rarity of enrolled specimens in linear clusters would support the hypothesis of very sudden death.

Collective Behavior

We propose that the linear clusters formed by *Ampyx priscus* result from the coordinated gathering and locomotion of individuals and therefore suggests a collective and synchronized behavior. We hypothesize that these trilobites moved in small groups on the seafloor, keeping a single-row formation by physical contacts via their long projecting spines and antennules and/or through chemical communication. Similarly, extant spiny lobsters perform mass single-file migrations by maintaining tactile contact between the tail fan of one individual and the [antennae] and tips of the anterior-most walking legs of its follower. Knowing that *Ampyx priscus* was blind, we hypothesize that mechanosensory stimulation via both genal and glabellar spines, or/and chemical cues, may have been a trigger that maintained group behavior.

Triggers, Functions and Benefits

Spawning congregations and synchronized moulting. Congregation of sexually mature individuals is frequent in [living arthropods] and is often related to reproduction or moulting. *Ampyx* may have performed comparable group migrations to distant spawning grounds during the reproductive season. This hypothesis is supported by the fact that *Ampyx* clusters almost exclusively consist of adult or sub-adult stages. Synchronized moulting is frequent in modern crustaceans and insects

[although] synchronized moulting, by definition, releases numerous exuviae which are absent from the *Ampyx* clusters.

Hydrodynamic cues. Field and laboratory studies on extant spiny lobsters have highlighted the possible relation between collective migrations and environmental disturbances. A drop in water temperature, higher water turbidity, and intense current induced by seasonal storms are assumed to be the main triggers for the mass migrations of these crustaceans, which always take place from highly disturbed shallow coastal areas to the edge of oceanic channels. The hypothesis that hydrodynamic cues may have driven a comparable behavior in *Ampyx* is realistic considering that these trilobites were potentially exposed to periodic environmental disturbances generated by storms. *Ampyx* is known to occur preferentially across the storm wave base and preferentially in the lower shoreface and upper offshore environments, suggesting possible migrations from storm-influenced to quieter and deeper areas. In summary, [these] two options can be seen as the most likely, [though they] are not mutually exclusive. *Ampyx* may have responded to environmental stress and reproduction signals by adopting the same behavior.

Origin of Collective Behavior

Ampyx shows that collective behavior in arthropods has [an ancestry that dates] to the lower Palaeozoic. This behavior was necessarily associated with a communication system between individuals involving motion and mechanical sensors, chemical signals, and possibly neurotransmitters. Although this behavior was not mediated by sight because *Ampyx* was blind, it implies neural complexity and the ability to process signals. *Ampyx* shows how a 480-million-year-old arthropod may have integrated its neural complexity into a temporary collective behavior related to seasonal reproduction or triggered by environmental cues. This behavior is likely to have been widespread among trilobites throughout the Palaeozoic. Collective behavior associated with communication and recognition systems probably evolved through natural selection as the Cambrian radiation proceeded and developed more extensively during the Great Ordovician Biodiversification Event when ecosystems became increasingly complex.

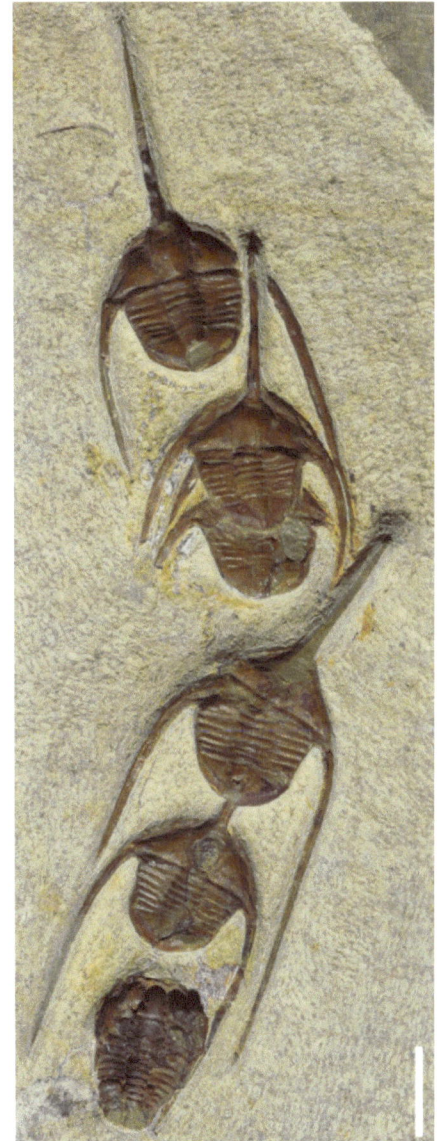

Edited excerpt and images from Vannier, J., Vidal, M., Marchant, R. et al. (2019). Collective Behavior in 480-Million-Year-Old Trilobite Arthropods from Morocco. *Scientific Reports, 9*(14941). Retrieved from https://doi.org/10.1038/s41598-019-51012-3.

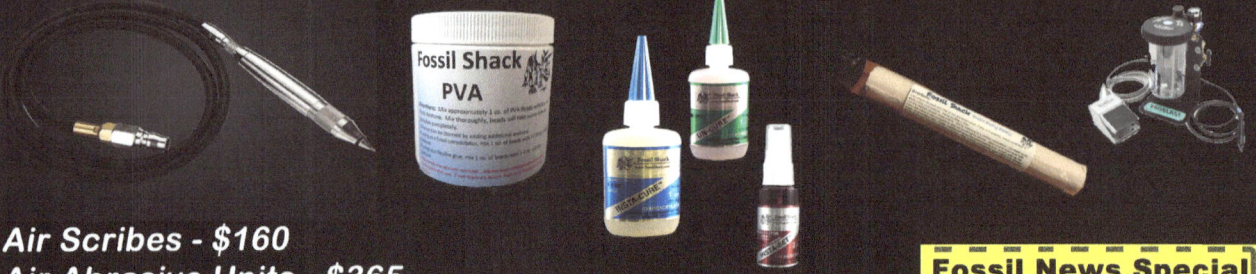

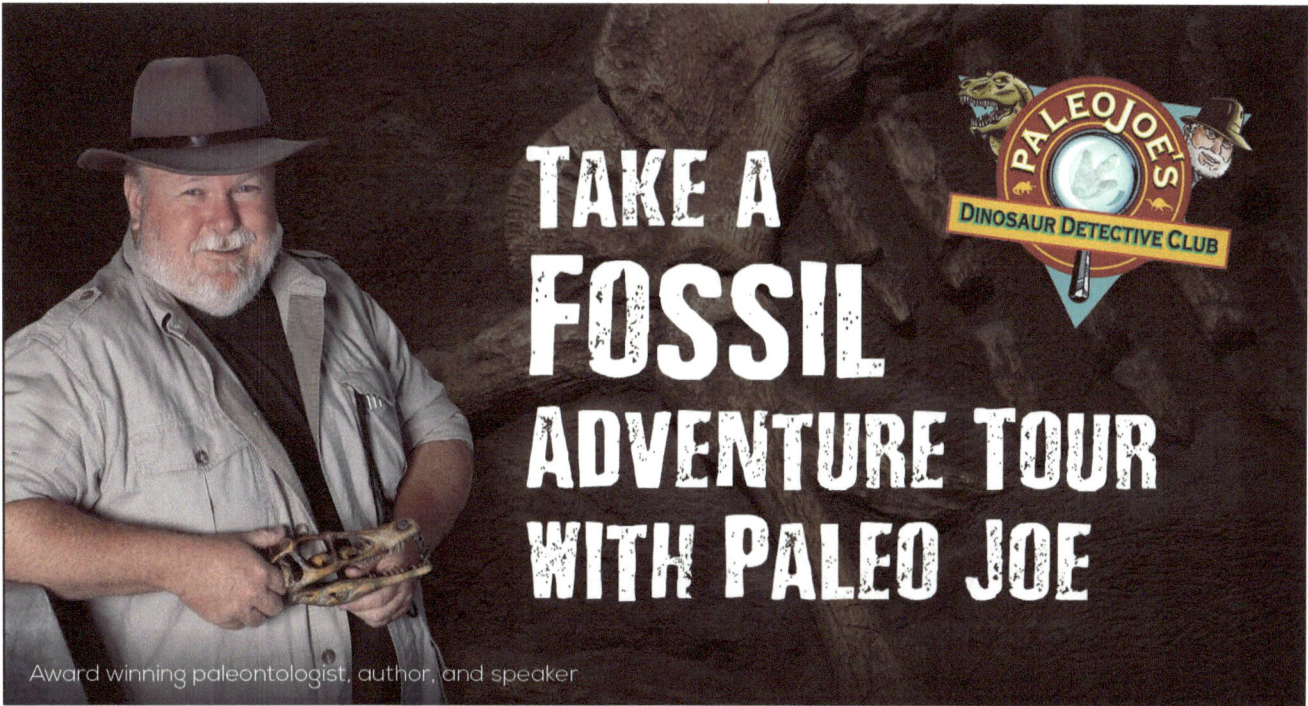

TAKE A FOSSIL ADVENTURE TOUR WITH PALEO JOE

Award winning paleontologist, author, and speaker

WYOMING'S FOSSILS & DINOSAUR TREASURES

Kemmerer & Thermopolis, WY

Package prices starting at

$799 Adult
(double occupancy)

$449 Children

Inclusions

- One night's accommodations in Evanston
- One night's accommodations in Riverton
- Breakfast at each hotel
- Two boxed lunches
- Two dinners
- Evening introduction and presentation with Paleo Joe
- Two full days of digging with Paleo Joe
- Admission to Fossil Safari
- Admission to Wyoming Dinosaur Center
- You GET TO KEEP some of the Treasures! (fish fossils only)

DIGGING IN MICHIGAN'S ANCIENT SEA

Alpena, Michigan

Package prices starting at

$549 Adult
(double occupancy)

$279 Children

Inclusions

- Two nights' accommodations at Thunder Bay Resort
- Two breakfasts at the resort
- Two boxed lunches
- One five-course dinner including wine pairing at Thunder Bay Resort
- Carriage and Elk ride
- One full day of digging with Paleo Joe
- Evening introduction and presentation by Paleo Joe
- Shipwreck glass-bottom cruise
- Admission to Thunder Bay National Marine Sanctuary - including Science on the Sphere
- You GET TO KEEP all the fossils you find!

CALL 313.575.8888 EXT. 185 OR 122 TO LEARN MORE!

EDUCATIONAL TOURS
BY CORPORATE TRAVEL

Check Your Paleo-IQ: What Was That Name Again?

Species of every imaginable animal, plant, fungus, or protist have been named to give credit to thousands of scientists and discoverers, but quite a few have also been christened in honor of (and sometimes for reasons other than honor) real-life musicians, movie stars, TV celebrities, and writers, both obscure and well known. In this issue, for example, paleontologist Mats Eriksson wrote about a new Devonian marine worm that he named for famed metalhead bass player, Alex Webster. Others have even been named for fictional characters.

A few folks seem to have earned more than their share of namesakes (Sir David Attenborough, the British broadcaster and natural historian, for example), and more famous names seem to crop up in a few groups of organisms more than others (trilobites probably win that prize). See if you can guess the famous person behind the name (most of them won't take long, but a few are puzzlers), and then try to imagine what kind of creature the name is attached to.

Torvosaurus gurneyi	*Kootenichela deppi*
Avalanchurus lennoni and *Avalanchurus starri*	*Mackenziurus johnnyi*
Isbergia planifrons and *Warburgia crassa*	*Leninia*
Bambiraptor	*Australopithecus afarensis*
Darwinius masillae	*Megalonyx jeffersonii*
Effigia okeeffeae	*Montypythonoides*
Chrichtonsaurus	*Obamadon gracillis*
Gagadon minimonstrum	*Electrotettix attenboroughi*
Han solo	*Amaurotoma zappa*
Scipionyx samniticus	*Sauroniops*
Trierarchuncus	*Mesoparapylocheles michaeljacksoni*
Jaggermeryx	*Confuciusornis sanctus*
Jenghizkhan bataar	*Anisonchus cophater*
	Xenokeryx amidalae

Kootenichela deppi

Kootenichela deppi: ...esting reason: It has large claws with elongated spines that reminded the paleontologist who described it of Edward Scissorhands, the character Depp played in the 1990 movie of the same name.

Mackenziurus johnnyi: Another Silurian phacopid trilobite, *Mackenziurus* honors four members of the American punk band, The Ramones, active from 1974–1996. *M. johnnyi* was named for Johnny Ramone, *M. joeyi* for Joey Ramone, *M. deedeei* for Dee Dee Ramone, and *M. ceejayi* for C. J. Ramone.

Leninia: An extinct Cretaceous ichthyosaur, *Leninia* was named for Vladimir Lenin, the head of the Soviet Russia from 1917 to 1924 and of the Soviet Union from 1922 to 1924.

Australopithecus afarensis: A trick question. The actual scientific name, *A. afarensis* doesn't have anything to do with a famous individual, but you may recall the nickname by which this proto-human was known in the media after her discovery in 1974: "Lucy." And "Lucy," of course, is a reference to the Beatles' song *Lucy in the Sky with Diamonds*, which, according to the legend, was played often and loudly in the Ethiopian expedition camp.

Megalonyx jeffersonii: A large, extinct Pleistocene ground sloth, named for the third president of the United States, Thomas Jefferson, who was an avid amateur paleontologist. French zoologist Anselme Desmarest christened *M. jeffersonii* in 1822 and, exactly 100 years later, Jefferson's name was used once more for the North American mammoth, *Mammuthus jeffersonii* (*M. jeffersonii* is now largely considered a synonym for other *Mammuthus* species). Jefferson has at least one more fossil to his naming credit: the large and striking bivalve, *Chesapecten jeffersonius*, believed to be the first North American fossil to be described in official scientific literature (1687). This Pliocene mollusk is the state fossil of Virginia.

Torvosaurus gurneyi: Fans of the Dinotopia book series may recognize this homage to illustrator and paleoartist, James Gurney, whose name is forever attached to this Late Jurassic megalosaur.

Avalanchurus lennoni and **Avalanchurus starri**: These Silurian phacopid trilobites are named after two of The Beatles, John Lennon and Ringo Starr. Two other species of **Avalanchurus** are named for the 1960s musical duo, Simon & Garfunkel: A. simoni and A. garfunkeli. Cousins of **Avalanchurus**, **Struszia harrisoni** and **Struszia mccartneyi**, are named for the other two Beatles: Paul McCartney and George Harrison.

Isbergia planifrons and **Warburgia crassa**: As the story goes, Swedish paleontologists Elsa Warburg and Orvar Isberg chose these less than flattering species names with each other in mind: Warburg christened **I. planifrons**, an Ordovician trilobite, with a species name that means "flat-headed"—or stupid—in Swedish. Isberg gave the name **W. crassa** to an Ordovician bivalve, knowing that "crassa" meant "thick"—as in fat.

Bambiraptor: A Late Cretaceous, bird-like dromaeosaurid dinosaur that was named after the Disney deer because of its small size.

Darwinius masillae: **Darwinius** is a basal, lemur-like primate from the Middle Eocene, and its genus name honors Charles Darwin on the occasion of the bicentenary of his birth.

Effigia okeeffae: **E. okeeffae** was an Upper Triassic archosaur that lived in what is now New Mexico. It was named after the painter Georgia O'Keeffe, who spent many years at Ghost Ranch, near where the specimen was found.

Chirichtonsaurus: **Chirichtonsaurus** was an ankylosaur that lived in China during the Late Cretaceous, and its name honors Michael Crichton, the author of Jurassic Park (most of the dinosaurs in Jurassic Park are Cretaceous species). **Cedrorestes crichtoni**, an Early Cretaceous iguanodon, is also named for the author, and another Chinese ankylosaur was named with **Tianchisaurus nedegoapeferima**, whose species name is formed from the surnames of the film's stars: Sam Neill, Laura Dern, Jeff Goldblum, Richard Attenborough, Bob Peck, Martin Ferrero, Ariana Richards, and Joseph Mazzello.

Gagadon minimonstrum: The "Lady Gaga-toothed little monster" was described in 2014 on the basis of a fossil jaw and some unusual associated teeth. **Gagadon** was a hoofed mammal that lived in what is now Wyoming.

Han solo: When the description of this small agnostid Middle Ordovician trilobite from China was published in 2004, its author said it was named "Han" for the Han people of China and "solo" because it was the only member of its genus. Later, though, he admitted that his friends had dared him to name a species after a Star Wars character.

Scipionyx samniticus: **Scipionyx**, known from a single specimen found north of Naples in 1981, was the first dinosaur known from Italy and is named for two people: Scipione Breislak, an eighteenth-century Italian geologist, and Scipio Africanus, the third-century BCE Roman military strategist and general who defeated Hannibal.

Trierarchuncus: You might consider this one a bit of a cheat. **Trierarchuncus** was a small, long-legged, bird-like but flightless dinosaur. It had a single large claw on each hand and, thus, its name means "Captain Hook." **Trierarchuncus** isn't really named for the Peter Pan character, though. Its name comes from the combination of "triarch" (the captain of a trireme in classical Greece) and uncus, meaning "hook" in Latin. Its specific name, **prairiensis**, "from the prairie," refers to the plains of eastern Montana where its fossils were first discovered.

Jaggermeryx: An extinct genus of semiaquatic anthracothere (ungulates related to hippopotamuses) from the Early Miocene of Egypt, **Jaggermeryx** was named for Rolling Stones front man, Mick Jagger, in part because it was believed to have large, fleshy lips. Jagger also has a trilobite to his credit (**Aegrotocatellus jaggeri**) and a tiny Permian marine gastropod, (**Anomphalus jaggerus**).

Jenghizkhan bataar: During several years in which the classification of the tyrannosaurid Mongolian dinosaur, **Tarbosaurus bataar**, was a matter of debate, the American writer and paleontologist, George Olshevsky, proposed the genus name **Jenghizkhan** in honor of Genghis Khan, the twelfth-century Emperor of the Mongol Empire. In fact, Olshevsky recognized three distinct genera of Late Cretaceous Mongolia "meat-osaurs," but, over the last twenty years or so, they have been revised into a single accepted genus: **Tarbosaurus**.

Kootenichela deppi: This extinct arthropod, a distant ancestor of lobsters and scorpions, was named for actor Johnny Depp, but for an inter-

Darwinius masillae, reconstruction © Nobu Tamura, CC License 3.0.

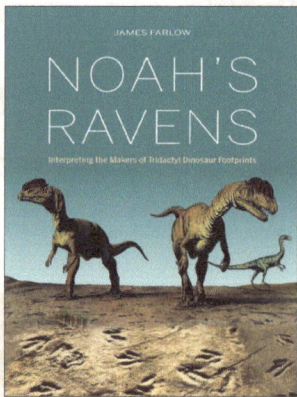

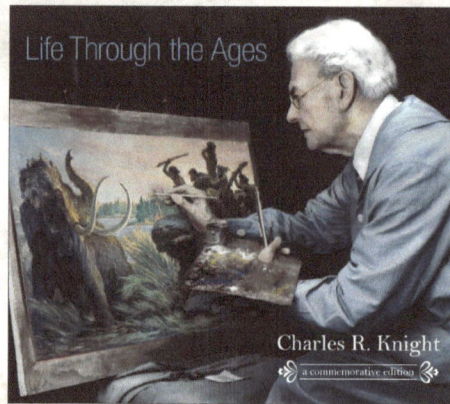

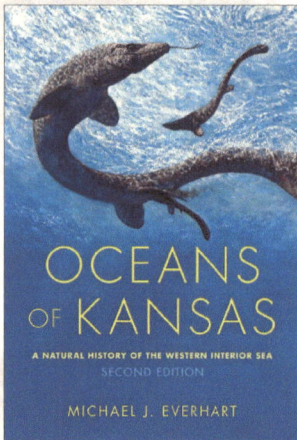

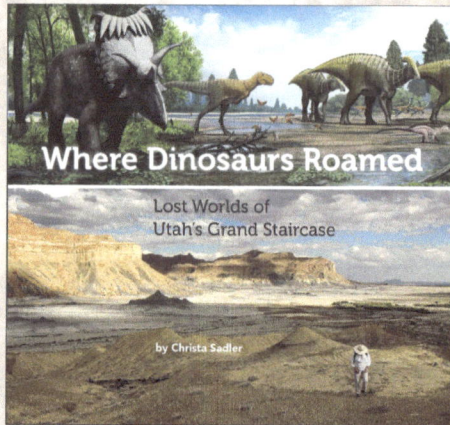

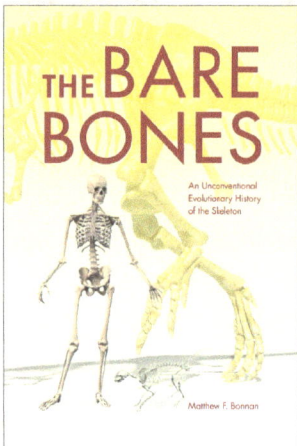

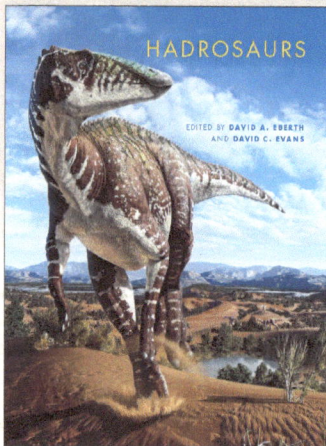

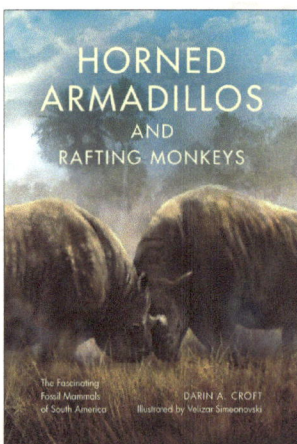

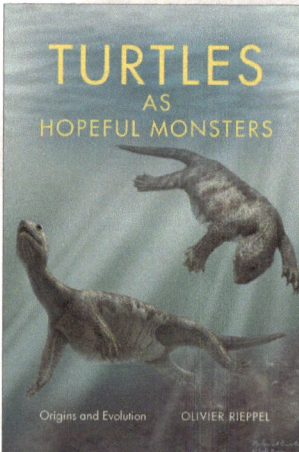

Montypythonoides: It's a shame this genus name for an extinct python didn't stick (it's now known as *Morelia*), because naming it for the comedy troupe Monty Python was just too perfect!

Obamadon gracilis: A small, extinct insect-eating Late Cretaceous lizard known from the Hell Creek and Lance Formations, was named for the forty-fourth president of the United States, Barack Obama. He also had an Ediacaran fossil, *Obamus coronatus*, named for him in 2018 because of the similarity of the creature's shape to Obama's ears.

Electrotettix attenboroughi: Found in amber from the Dominican Republic, this extinct pygmy locust is only one of the fossil (and other) species named for Sir David Attenborough. He also inspired the names for *Materpiscis attenboroughi*, a Late Devonian armored fish from Australia; *Mesosticta davidattenboroughi*, a Cretaceous damselfly; *Microleo attenboroughi*, a miniature marsupial lion from the Early Miocene; and *Attenborosaurus*, an Early Jurassic plesiosaur from England.

Amauritoma zappa: This Permian gastropod from Nevada was named for Frank Zappa (1940-1993), the genre-bending American musician, composer, and musical satirist. Zappa also lent his name to an Early Miocene gerbil-like rodent, *Vallaris zappai*.

Saurniops: The genus name of this carnivorous carcharodontosaurid dinosaur from the Late Cretaceous of Morocco means "The Eye of Sauron" and is, of course, a reference to the terrible mountain entity from Tolkien's *Lord of the Rings.*

Mesoparapylocheles michaeljacksoni: This extinct Cretaceous hermit crab species was named for pop singer Michael Jackson because paleontologists discovered it on the same day they learned of Michael Jackson's death in 2009.

Confuciusornis sanctus: This early bird relative from the Early Cretaceous of China, and an ornithischian dinosaur, *Tianyulong confuciusi* from the Early Cretaceous Jehol group in Western Liaoning Province, China, were both named in honor of the Chinese philosopher, Confucius, who lived between 551 BCE and 479 BCE.

Anisonchus cophater: This story comes from the life of Edward Drinker Cope that American paleontologist, geologist, and eugenicist, Henry Fairfield Osborn, published in 1931. In it, Osborn quoted a letter from Cope in which Cope explained that he had given the placental Palaeocene mammal, *Anisonchus*, the specific name "cophater" for "all the Cope-haters who surround me." Marsh, in his turn, named a marine reptile after me." Marsh: *Mosasaurus copeanus*. A hundred years later, an American evolutionary biologist had the final word: he named yet another primitive Palaeocene ungulate *Oxyacodon marshater.*

Xenokeryx amidalae: Fossils of this ancient ancestor of the giraffe from Spain were first described in 2015. The leader of the team who discovered the new species said they chose the name "amidalae" in honor of the *Star Wars* character Padmé Amidala Naberrie because its horns resembled the elaborate hairstyles she wore as queen of Naboo.

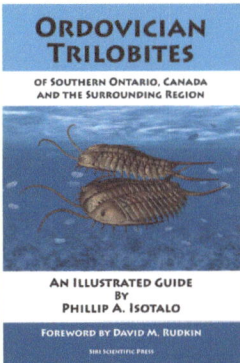

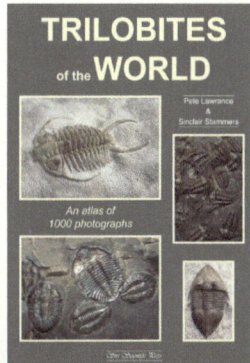

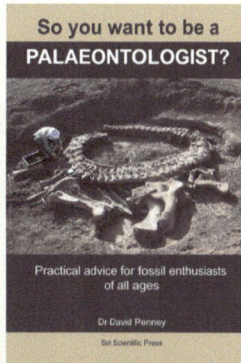

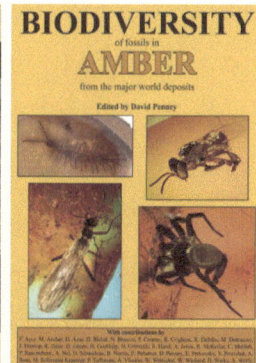